Climate Change is an Opportunity

Why We Need Principled Capitalism

David Blockley
Emeritus Professor of Engineering
University of Bristol
UK

CRC Press
Taylor & Francis Group
Boca Raton London New York

CRC Press is an imprint of the
Taylor & Francis Group, an **informa** business
A SCIENCE PUBLISHERS BOOK

***Cover credit*:** Image by David Blockley and via Pixabay

First edition published 2024
by CRC Press
2385 NW Executive Center Drive, Suite 320, Boca Raton FL 33431

and by CRC Press
4 Park Square, Milton Park, Abingdon, Oxon, OX14 4RN

CRC Press is an imprint of Taylor & Francis Group, LLC

Library of Congress Cataloging-in-Publication Data (applied for)

ISBN: 978-1-032-62941-4 (hbk)
ISBN: 978-1-032-62942-1 (pbk)
ISBN: 978-1-032-62944-5 (ebk)

DOI: 10.1201/9781032629445

Typeset in Times New Roman
by Shubham Creation

Remarks by President Biden on Supporting Ukraine, Defending Democratic Values, and Taking Action to Address Global Challenges | Vilnius, Lithuania | The White House | July 12th, 2023.

"*My friends, at the most fundamental level, we face a choice—it's not a hyperbole—we face a choice: a choice between a world defined by coercion and exploitation, where might makes right, or a world where we recognize that our own success is bound to the success of others.*

When others do better, we do better as well—where we understand that the challenges we face today, from the existential threat of climate change to building a global economy where no one gets left behind, are too great for any one nation to solve on their own, and that to achieve our goals and meet the challenges of this age, we have to work together.

And I mean this sincerely: The world is changing. We have a chance to change the dynamic.

You know, we all must summon the common will to—to actually address the existential threat of accelerating climate change. It's real It's serious. We don't have a lot of time. It is the—the single greatest threat to humanity.

And it's only by working together that we'll prevent the worst consequences of climate change from ravaging our future and that of our children and grandchildren.

We also have to recognize our shared responsibility to help unlock the enormous potential that exists in low- and middle-income countries around the world—not out of charity, because it's in our own self-interest. We all benefit when more partners stand together, working toward shared goals. We all benefit when people are healthier and more prosperous. And that's not, again, hyperbole. It's true. We all benefit when more entrepreneurs and innovators are able to pursue their dreams for a better tomorrow.

You know, so we need to update our toolset to better address the needs of today in this interconnected world. A world where climate disasters, pandemics, conflicts spill over borders and make it harder to address the challenges of poverty and instability that hold so many people back."

Prologue

A very poor farmer was struggling to feed his family. He prayed constantly for help. One night his prayers were answered in a dream. An old woman told him to go to his local market the next morning and buy the first goose he saw. The old woman said that ***as long as he wasn't greedy****, the goose would live forever and lay one golden egg each day. So, the farmer went to the market and bought the first goose he saw. The very next morning the goose laid a golden egg and continued to do so every day after. The farmer and his family became wealthy. But the farmer was jealous of the much wealthier village banker. He was so jealous that he forced the goose to lay two golden eggs a day, and then three. The goose soon died of exhaustion.*

Life on earth is the golden egg begat by the goose of planet earth. We are the farmer and the village banker. The goose is exhausted and may soon stop laying eggs. It may even die. Life on earth will be no more unless we nurse the goose back into better health.

HOW DOES CLIMATE CHANGE PRESENT US WITH AN OPPORTUNITY?

The simple answer is that we have an imperative, as never before, to change our ways. The entire human race is facing its greatest ever existential challenge. I wrote this book because I feel a growing sense of looming disaster. We are not nursing the golden goose as we should. Yes, we are making some progress through initiatives such as the 'Green New Deal (GND)' and the USA Inflation Reduction Act—so it's not all gloom[1]. The Intergovernmental Panel on Climate Change (IPCC)[2] continues to provide important rigorous and balanced scientific information based on published literature. The UK HM Treasury Net Zero Review Interim report of December 2020[3] states, '*Climate change is an existential threat to humanity. Without global action to limit greenhouse gas emissions, the climate will change catastrophically with almost unimaginable consequences for societies across the world*'. The UK and 77 other countries are committed to reaching 'net zero' greenhouse gas emissions by 2050[4]. But past agreements such as the Tokyo Protocol and Paris[5] are not delivering as William Nordhaus has pointed out[6]. Policies such as carbon taxes[7] or cap and trade[8] are fragmented and because they are voluntary with no penalties for not taking part they suffer from the 'Free Rider Problem'[9]. Joyeeta Gupta of the Amsterdam Institute for Social Science Research says[10] "*something has to happen drastically in the next two to three years*" if the world is to limit temperature rises to 1.5C. '*Everyone has been delaying action... Now we have to go down so fast [in terms of emissions] that it is very painful.*'

The 2022 IPCC report[11] says that '*nature provides major tools to put us back on track. The future we fear is not inevitable. Oceans, forests, and other ecosystems already absorb and store about half of global carbon emissions. The despair we feel from climate projections must turn into action. Solving climate change is an opportunity to tackle problems we have struggled with for generations. Change won't be easy, but we have no other choice. Either we allow our planet to be destroyed, or we fight—clear-eyed—for a better world*'. The 2023 IPCC report[12] states '*There is sufficient global capital and liquidity to close global investment gaps...but there are barriers to redirect capital to climate action both within and outside the global financial sector...Reducing financing barriers...would require clear...support by governments, including a stronger alignment of public finances...improving the risk-return profile of investments. ... financial actors, including investors, financial intermediaries, central banks, and financial regulators can shift the systemic under-pricing of climate-related risks*'. This latest announcement triggered the UN secretary Antonio Guterres to say[13] '*This report is a clarion call to massively fast-track climate efforts by every country and every sector and on every timeframe. Our world needs climate action on all fronts: everything, everywhere, all at once*'. British politician Dame Margaret Beckett, chair of the Joint Committee on the UK National Security Strategy accused the UK government of choosing to delay the introduction of new regulations[14]. She said, '*This is akin to organising a fire safety awareness course to respond to a blazing fire*'.

Martin Wolf[15] writes '*Pessimism about humanity's ability to address climate change is understandable. Time is limited, talk plentiful and action negligible. But we can only start from agreement that there is indeed a threat worth addressing. That may now be emerging, even in the US. Turning such a consensus into a workable, globally replicable, and politically acceptable plan is going to be very hard. But despair is not an option*'.

The Mayor of Bristol, the city in the UK where I live, says[16] '*We need to reframe the way we think about the economy...we should be saying to the world and to the markets...the kind of country we are going to be in 2027 is resilient in mental and physical health, the capacity of our children to learn, the physical infrastructure we have in place, the stock of affordable houses and social stability...We should classify public spending as public investment*'.

Venture capitalist John Doerr writes[17] '*we're not cutting our emissions fast enough...unless we course correct with urgent speed...we'll be staring at a doomsday scenario*'. He has suggested 10 key objectives to cut greenhouse gas emissions to net zero by 2050. They are electrifying transportation, decarbonising the (energy) grid, fixing food, protecting nature, cleaning up industry, removing carbon, winning politics, and policy, turning movements into action, innovating, and investing. His website is tracking progress on each target and is showing that we are way off course in most of our attempts to reach his targets.

We know what we need to do. The problem is doing it. How do we convince our politicians that more urgency is needed? Doerr's final target of getting sufficient investment is the key. The net result is that governments and industry are not doing enough to care for the goose. They are reacting too slowly. As we shall see throughout this book, it is fear of national debt that is one of the key factors in holding us back.

Nevertheless, Yeva Nersisyan and L Randall Wray, and separately Richard Murphy, are showing us how we can pay for the Green Deal[18]. Nersisyan argues that the Green New Deal is MEOW (Moral Equivalent of War). The basis of their case is that affordability cannot be an issue for the sovereign USA government (or any sovereign state with fiat currency—see later discussion on Modern Monetary Theory MMT). They write '*The costs of extinction of the human species—from the point of view of humans, at least—is beyond measure. Even if we calculate the costs of the GND as $93 trillion (as one hysterical estimate puts it) over the next decade, that is puny in comparison with the discounted cost of total destruction of human life on planet Earth...The future of humanity lies in the balance. Half measures will not do*'. The problem, they say, will be inflation if sufficient resources cannot be diverted. If inflation is likely, we need to put in place anti-inflationary measures, such as well-targeted taxes, wage, and price controls, rationing, and voluntary saving. They say that the USA already has a financial system to handle the costs, but the problem is whether we can mobilize the required resources. Murphy[18] has proposed green quantitative easing to kick start economic activity to deliver a new green deal.

A panel of experts working for the COP27 meeting in November 2022 have written[19] '*Acting on climate is about transforming our economies, particularly our energy systems, through investing in net zero, adaptation, resilience, and* *natural capital requiring strong investment and innovation...The world needs a breakthrough and a new roadmap on climate finance that can mobilise the $1 trillion per year in external finance that will be needed by 2030 for emerging markets and developing countries...A major, rapid and sustained investment push is needed to drive a strong and sustainable recovery out of current and recent crises, transform economic growth...The key investment priorities must encompass transformation of the energy system...Country/sector platforms driven by countries can bring together key stakeholders around a purposeful strategy, scaling up investments...The scale of the investments needed...will require a debt and financing strategy that tackles festering debt difficulties*'.

In March 2023, the UK government and the Bank of England published four major policy statements[20]. The strategy is to mobilise private-sector money for investments in green industry with consultations on carbon border taxes. Fran Boait of Positive Money welcomed the commitment to map and track investment but fears that capital will still not be generated at the speed required to keep temperatures to 1.5°C. Levels of public investment in net zero are still falling far short of what is required. The Bank of England is reviewing regulatory policies but should be targeting lending schemes. A continued use of fossil fuels is justified by an investment in carbon capture under the North Sea.

Across the world countries like Russia have been slow to accept there is a problem[21]. But wildfires have consumed huge stretches of Siberian Forest and flooding has disrupted farming which, combined with droughts, has threatened Russia's food security. Russia has pledged to reach net zero by 2060 but has not yet set out how they plan to do it. Clearly the war in Ukraine has complicated the issue. We in the West need to reach out to the Russian people who, if properly informed, would almost certainly want action on climate change rather than war. China has also promised action. President Xi of China said in a speech to the 75th Session of the UN General Assembly September 2020 that[22] '*Humankind can no longer afford to ignore the repeated warnings of nature...We aim to have CO_2 emissions peak before 2030 and achieve carbon neutrality before 2060. We call on all countries to pursue innovative, coordinated, green and open development for all, ...achieve a green recovery of the world economy*'. Some will dismiss this as propaganda. Covid-19 has made the picture more problematic, but China must be involved in global action since China is responsible for nearly one third of global greenhouse gas emissions.

THE SIGNS ARE THERE

Three in four people across the 27 countries think that our world is more divided than ever[23]. It matters. The greatest tensions are between people with different political views, religious beliefs, the rich and poor, levels of corruption and between immigrants and people born in the country. Global terrorism and our response to it frightens us.

For these many reasons we need a deep rethink of how we behave towards each other and to the planet—a change of collective perspective, mindset, social norms, and culture. Leading democracies are increasingly polarised in pernicious entrenched animosity and intolerance[24]. The demagogues of totalitarian nations either threaten or wage war (such as the Russian invasion of Ukraine and the posturing of China over Taiwan) and extreme religious terrorists continue their campaigns against anyone who doesn't agree with them[25].

Everything seems to be in an unstable state of flux. Uncertainty is rife. We just don't know what's around the corner. People are anxious. Mental health problems after the Covid-19 pandemic at an all-time high amongst the under 18s[26]. Understandably many of them are angry at the generations that created the problem. There has been a massive decline in trust of public institutions from the news media to politicians to 'experts' of any kind[27].

Even the most narcissistic vindictive dictator or self-centred international greedy criminal can read the physical signs. Melting ice caps and permafrost, extreme cold and deep snow, rising sea levels, excessive rainfall, storms, flooding, hurricane winds together with extreme heat and drought have no regard for national boundaries or lifestyle. Food and energy supplies are already being disrupted and it will get worse. If a dictator does not realise that his power is threatened if he does not act to mitigate these kinds of disasters, then eventually the people will overthrow him. But then it may be too late. Even the most anti-social, hardened criminal will eventually have to co-operate to 'save the skin' of the criminal network on which he depends.

Of course, it is unrealistic to think we won't continue to be divided and criminals won't continue to be corrupt. But climate change is the one issue we cannot afford division. We have to agree on what needs to be done because if we don't, there will be even more suffering and life as we know it may well perish. There will be nothing left for the dictator to dominate and the criminal to steal from.

AN IMPOSSIBLE TASK?

The task seems impossible—even utopian to contemplate. But we can and must make it possible. It's doable because we have the technical capability and organisations like the IPCC in place to make it happen. The political will is there but it is weak. To reinforce the point made earlier, the UK Climate Change Committee report to Parliament 2022[28] says that progress is lagging the policy ambition. '*In targets, the UK is indeed a world leader...despite important achievements in renewable energy and electric vehicles, the Government is failing in much of its implementation. Sharply rising fuel costs should have given added impetus to improving energy efficiency, yet the necessary programmes are not in place. We are still building new homes that do not meet minimum standards of efficiency and will require significant retrofitting.*' After a legal challenge from environmental groups Mr. Justice Holgate said, in judgement, that the UK government strategy had failed to meet its obligations under the UK Climate Change Act 2008[29].

The pressures on politicians to act on short term issues, together with the election cycle, mean that they tend to neglect the long term[30]. But we need them to not only think long term but to act on it–the alternative is too awful to contemplate. As I said earlier, and will discuss later, they must rethink their fear of government debt[31] and invest much more money into preparing our joint future. Once we have that message on board then the imperative becomes obvious. All else pails in significance.

In the three parts of this book, I will set out the three sets of changes of behaviour that I think we need to make and five axioms on which we can base those changes. In sum they are thinking differently (about ourselves and our systems), testing diligently (the fidelity of what we are told) and acting effectively (towards a common purpose). The axioms concern purpose, layers of modelling of parts and wholes, interdependence, change and learning.

In Part 1, I maintain that we become what we *think*. That is why, collectively, we need to think differently to change social norms. Our priorities should be to rethink six things. First, who we are and what is our place in a deeply divided world. I will argue that we need to see ourselves as citizens, not just of our own countries, but also of the whole world. Second, how we relate to nature and our environment. I think the lifestyle of many insulates them from nature. They would benefit from some kind of direct daily experience with nature. Third, how we identify and manage systems of both the physical and social kind. Many of our systems are poorly integrated and not 'joined-up'. Fourth, what should be the roles of engineering, and economics. I will describe how they are bedfellows. Fifth, why we need to change democratic capitalism to principled capitalism—a new form of capitalism where we all recognise our obligations to each other and to the natural world based on moral virtues.

I will also introduce my first three axioms (of five) which help us think about the explicit challenge of the complexity of our systems. The complexity and intangibility of today's businesses creates an incredibly large information gap between people and organisations. As religiosity in the West declines, the faith gap between that which we can trust to give our lives meaning and our fears, has widened. We have lost the power of the 'golden rule' shared by most religions—we should not do to others that which we would not have them do to us. In other words, we should treat others as one would like to be treated[32]. For believers God is the imperative that gives meaning to that rule. We will be punished in hell for our sins and rewarded in heaven for our goodness. God is also how many explain that which cannot be comprehended by the human mind—such as the suffering of innocents. Extreme interpretations of the religious message have created, for example, the 'Christian Right' as political factions in the USA and the Islamic terrorists of Jihadism. Many reasonable nonreligious people may not have the golden rule uppermost in their minds, but most have an innate sense of fairness and tolerance that derives from it. Nevertheless, we seem to have lost it in our everyday behaviour. Common examples are when we fail to show gratitude, lack respect (especially on social media), racial hatred and misogyny, gossip behind someone's back, don't play by the rules (not pulling our weight in a team), win at the expense of another (cheat), make biased judgements (are unfair), criticise unfairly (intolerance). Those who have lost faith in God may simply put their faith in family, friends and those professionals who do their best to help others such as doctors, pharmacists, firemen, search and rescue teams, pilots, engineers and even strangers who seem genuine and honest. Faith means trusting in something you cannot prove. Faith implies trust. But actions speak louder than words. Trust is earned by being sincere, loyal, reliable, honest, transparent, kindly, tolerant, and respectful of the views of others. Trust is earned by being good parents, children, and friends who genuinely show regard, concern, love, and help for one another. But trust is fragile and easily broken and not easily repaired as we shall see.

Dictators tend to be insecure precisely because they don't know who they can trust—but inevitably they have to trust something (for example, violent repression of the population) or someone (their accomplices in power). As well as controlling the population by force they will, more subtly, control the flow of information—especially the news media and the internet. They control rivals by rewarding them or eliminating them. They cling onto power because they know that if they lose it, they will probably be jailed, exiled, or killed. Their power comes at the expense of goodwill and peace of mind. Criminals, like dictators, have to trust in their network of fellow gangsters—honour amongst thieves. By its very nature, their power is fragile in a constant battle to be the 'top dog'.

Global citizenship must entail every kind of faith in a global 'mix' which contains, at its heart a faith in common humanity[33] and our natural world. If it is to work then it has to include all forms of national governance—liberal democracies, autocracies, and all those in between.

There is good and bad in everyone—but most people do care about others—even in autocratic states. For example, we need to reach out to the Russian people and avoid demonizing them for the acts of their insecure rulers. Like those of us in the West they want to be in control of their lives. At home acts of individual cruelty, such as knife stabbings by teenagers, horrify most of us but they are all too commonplace in democratic modern life.

The current constant diet of bad news reported by the media is swamping much of the good in some people. They get anxious, often depressed, and consequently suffer from poor mental health. Unfortunately, some politicians exploit these weaknesses. In Russia the state media blame the 'wicked' west who want to take over Russia. In western democracies some politicians purposely bring out the bad in people by appealing to our fears and give us permission to blame others unlike ourselves. This kind of debasement leads many to wholesale cynicism, and big divisions between 'them' and 'us'. Unfortunately, these messages convey the idea that 'they', the elite, are out for themselves and are not interested in helping 'us', the workers, who are falling behind. This contributes to a loss of trust in democratic politicians that is widespread[34] and a threat to democracy as Martin Wolf has described very clearly[35].

The solution lies in finding ways of harnessing that inner good in all of us. We are born innocent—it is life's experience that makes imperfect. Some people have, perhaps genetically, an inbuilt tendency to be selfish. They are unable to see the world through the 'eyes' of others and hence to consider the interests of those others. We have to help such people not put them in prison—unless of course they are a real danger to society. They should be encouraged and assisted to see the world as others do and helped to see the effects of others life's experiences. If they have harmed someone then they should be counselled into understanding the consequences of what they have done. For example, restorative justice brings those harmed by crime or conflict and those responsible for the harm together to try to repair some of the harm and find a positive way forward. We should provide support to help them control their emotions, such as anger, and encourage tolerance. Of course, some people are simply ignorant, perhaps because life's raw deal has limited their exposure to any wider views. We have, as citizens of the world, a responsibility to educate them and help them to develop learning power[36]. Learning power is the capacity to use reflection, self-awareness, and relationships with others to learn more effectively. Some people are corrupt—for many complex reasons including that they are selfish and don't want to work, or they become addicted to drugs, or have a difficult upbringing, or get into debt through gambling or bad business deals. We need to publish more stories about the many acts of good that occur every day and encourage people to find a new sense of purpose and self-fulfilment by doing good things rather than bad. We need the 'good' to become a major part of our social norms—the unwritten rules of beliefs, attitudes and behaviours that form a dominant part of our culture.

Donations to charitable and emergency funds show that people are distressed by unfairness and suffering. Witness the large sums donated by the public after the 2023 earthquake in Turkey and Syria and the war in Ukraine. Ordinary people do not want war, even in autocracies. They want to live in peace. It is governments who persuade and require them to fight to gain territory, revenge, revolution or for economic gain, or religious or nationalist differences. People value tolerance and decency—it is some large organisations and corporations that put profit for shareholders before the well-being of individuals and who exploit the resources of the natural environment.

Social norms do change. Examples are the decline in smoking in the UK over my lifetime, the changing stigma of drinking and driving and the widespread use of seat safety belts. Of course, there will always be people who resist but social norms can be reshaped as Cristina Bicchieri asserts[37]. To change a negative norm, she says we must change people's expectations with open conversations.

In sum the central message of Part 1 is that by renewing our thinking and by becoming citizens of the whole world, we can bring about a new form of principled capitalism with obligations based on moral virtues.

In Part 2, I argue that misinformation and social manipulation or grooming have become major problems. Climate deniers cling to conspiracy theories or simple deny the evidence as being inconclusive since change is part of the natural cycle. Other deniers say that climate change is simply too expensive to fix or that warmer summers will make farming more productive. Some say that we should not damage our own economy because other countries are not doing what they should—there are bigger problems closer to home. Other different examples of social manipulation include persuading teenage children to share intimate details of themselves and even to commit suicide. To minimise such tragedies, we must learn how better to ***test*** information before we act on it. That requires a deeper understanding of scientific evidence and uncertainty.

I will introduce axiom 4 concerning the ubiquity of change. The internet has transformed how we communicate—the widespread use of the world wide web is relatively recent, but its effects are profound. It has made possible the sharing of excellent information about specialist topics such as health, economics, and technology. It has transformed the publication of research. But some of the material is unreliable and simply false. Future changes will be disruptive as we move to new forms of transport and information technologies such as artificial intelligence (AI) and robotics take hold. It will become even more important that we become more discerning before we act on what we are told. We have to learn how to discriminate better the information 'wheat from the chaff'. This will become ever more critical as AI develops because it will present information some of which is spurious.

The current spread of fake news and the influence of social media is causing immense harm. We have to help people to reject rational ignorance (more on this later) criticise and *test* what they are told before they act. The complexity of systems interdependencies, and the lack of clarity on ethical issues are difficult to test adequately. Is there a 'right' thing to do in specific situation (for example, changing gender) and if so, can we test it out? Our present form of democratic or liberal capitalism encourages greed and corruption. Steven Pearlstein[38] thinks we have a kind of capitalism that corrodes social capital by undermining trust and discouraging cooperation. He writes '*we are now caught in one of those self-reinforcing, downward spirals in which government dysfunction, rising inequality, the erosion of social capital and slowing rates of economic growth are all feeding off each other, with more of one leading to more of all the others. Such vicious cycles are very hard to stop. The rise of the Tea Party and the election of Donald Trump (in the USA) are both consequences*'.

I think that we need to do at least four things. First, to do everything we can to help people to distinguish fake news and on-line gossip from bone-fide facts and well researched opinion including high quality journalism. Second, to help everyone to better understand the role of science in making policy. Third, to avoid governments making misleading statements such as 'We are following the science' by the UK government during the Covid-19 crisis. Fourth, to revalue how to discuss difficult issues critically with *tolerance* and goodwill and avoiding name calling and insulting behaviour that we see in our UK MPs across the floor of the House of Commons. We have to learn to agree to disagree amicably and certainly not kill each other merely because we have different opinions. We must reject rational ignorance where we repudiate relevant information that harms us either individually or collectively.

The central message of Part 2 is that we must encourage people to become more discerning about the information they are given and an ability to test it before acting on it.

In Part 3, I maintain that we build trust through the way we *act*. The core values of people are revealed by what they do—and it shows. Collectively our shared values have become increasingly blurred[39]. That is why we need to reevaluate our moral norms. We also have to 'get to grips' with uncertainty and risk. Uncertainty breeds anxiety and fear. But as we uncover the complexity of modern life, including who and what to trust, then it is something with which we have to accept and come to terms with. That is particularly so for unknown unknowns and surprises such as a pandemic. The success of reductionist science[40] has led us to focus on prediction. The complexity we are discovering in our world is requiring us to 'feel' our way through difficult situations. I introduce axiom 5 concerning complex systems that often cannot be 'solved' rather they have to be managed to desirable outcomes. We must 'learn how to learn' to find our way through uncertainty. That does not mean we abandon reductionist science, but it does mean changing emphasis. We need to follow Karl Popper's wise advice to advance 'piecemeal social engineering'. Finding our way towards a better future by reform and not by revolution. We need a much more nuanced understanding of the limitations of predictions. This is especially true in economics where predictions are an essential part of forming policy but are often wrong. There are good reasons why we can dependably predict physical events like the movements of the planets, but we can't predict who will win the war in Ukraine.

Finally, we need to recognise and nurture what Aristotle called *phronesis* or practical wisdom—the capacity for determining what is good for both the individual and the community, a virtue and a competence, an ability to deliberate rightly about what is good in general, about discerning and judging what is true and right. *Phronesis* is an idea that has got lost in the mists of history and it needs to be resurrected.

The central message of Part 3 is to pull together the ten pillars of principled capitalism. Don't do to others that you wouldn't have them do to you, become citizens of the world, nurture systems thinking and practical wisdom, apply modern monetary theory, invest in the public good, nurture ingenuity and innovation, tackle fake news, reduce inequality, embrace all forms of obligation, and root out corruption.

WE NEED A SENSE OF COMMON PURPOSE

It is unrealistic to think that we can suddenly make all of our conflicts and disagreements go away. Yet we know that the ventures we undertake are more likely to succeed if those taking part have a sense of common purpose. In truth we have two choices over the impending catastrophe posed by climate change. We can succeed by coming together with common purpose or we can fail by continuing to be fragmented and divided (the reality no doubt will be one of all shades in between). If we fail then, at worst life on earth will perish, and at best there will be damage, destruction, and loss of life on an unprecedented scale.

If we succeed, then multiple opportunities open up. New social norms will help us flourish. By being global citizens, we will require much better behaviour by our political leaders. That will give us a much firmer foundation on which to meet the other big challenges of our age. These include the crisis in democratic capitalism, identified so clearly by Martin Wolf[24], the scourge of extreme inequality, new forms of pandemic, our relationship with potentially disruptive engineering, and technology (including artificial intelligence), and the loss of biodiversity on both land and sea[41].

END NOTES

1. The term 'Green Deal' covers various green policies dating back to 2007. The Green New Deal was promoted by Alexandria Ocasio-Cortez and Sen. Ed Markey in the USA Congress in 2019. It failed to pass in the Senate. (See D'Souza (2022) Understanding the Green New Deal & What's in the Climate Proposal, Investopedia, https://www.investopedia.com/the-green-new-deal-explained-4588463). In 2022 US President Joe Biden signed 'The U.S. Inflation Reduction Act (IRA)' (see https://www.whitehouse.gov/wp-content/uploads/2022/12/Inflation-Reduction-Act-Guidebook.pdf). The aim is to reduce the use of fossil fuels and switch to clean energy in the USA. It commits to large tax credits over ten years. Many Europeans sees this as protectionist. In 2023 the EU announced its European Green Deal (see https://commission.europa.eu/strategy-and-policy/priorities-2019-2024/european-green-deal_en). This plan has no massive tax credits and no protectionism but higher carbon taxes. All 27 EU Member States committed to turning the EU into the first climate neutral continent by 2050. To get there, they pledged to reduce emissions by at least 55% by 2030, compared to 1990 levels. The Green New Deal in all countries requires public planning that is currently lacking. The state has a direct role but many dispute that especially in the USA. Increased public and private spending may lead to a sustained boom, that will need to be managed. There is a need to spend to meet human needs – if we don't then we will suffer severe consequences.
2. The Intergovernmental Panel on Climate Change. See https://www.ipcc.ch/
3. HM Treasury. 2020. Net Zero Review: Interim Report https://assets.publishing.service.gov.uk/government/uploads/system/uploads/attachment_data/file/1004025/210615_NZR_interim_report_Master_v4.pdf
4. HM Treasury. 2021. Net Zero Review: Analysis exploring the key issues https://assets.publishing.service.gov.uk/government/uploads/system/uploads/attachment_data/file/1026725/NZR_-_Final_Report_-_Published_version.pdf. See also https://sdg.iisd.org/news/77-countries-100-cities-commit-to-net-zero-carbon-emissions-by-2050-at-climate-summit/. See also McKinsey Sustainability (2021). Solving the net-zero equation: Nine requirements for a more orderly transition. https://www.mckinsey.com/capabilities/sustainability/our-insights/solving-the-net-zero-equation-nine-requirements-for-a-more-orderly-transition
5. European Commission. 2015. The Paris Protocol—a blueprint for tackling global climate change beyond 2020. See https://commission.europa.eu/publications/paris-protocol-blueprint-tackling-global-climate-change-beyond-2020_en
6. Nordhaus, W. 2020. Nobel laureate William Nordhaus, the economics of climate change. https://www.youtube.com/watch?v=5DG5i8BGaXo
7. Kagan, J. 2022. What is a Carbon Tax: Basics, Implementation, Offsets, Investopedia See https://www.investopedia.com/terms/c/carbon-dioxide-tax.asp
8. Kenton, W. 2020. Cap and Trade Basics: What It Is, How It Works, Pros & Cons. Investopedia. See https://www.investopedia.com/terms/c/cap-and-trade.asp

9. Investopedia Team. 2020. Free Rider Problem: Explanation, Causes, and Solutions. Investopedia. See https://www.investopedia.com/terms/f/free_rider_problem.asp
10. Mooney, A. 2013. Is 1.5C still realistic? The crumbling consensus over key climate target. Financial Times. March 16. See https://www.ft.com/content/450a59bb-7c83-4d04-851f-0bbc120c09f7?emailId=b86221d5-5084-48d9-8354-4f07c1b18bf8&segmentId=c393f5a6-b640-bff3-cc14-234d058790ed
11. Griscom, B. 2022. 5 takeaways from the latest IPCC report. See https://www.weforum.org/agenda/2022/04/ipcc-report-2022-key-takeaways/
12. IPCC (UN Intergovernmental Panel on Climate Change) Synthesis Report (AR6) 2013. Summary for Policy Makers. See https://report.ipcc.ch/ar6syr/pdf/IPCC_AR6_SYR_SPM.pdf.
13. Roston, E. 2023. World's top climate scientists issue blunt warning: Our chance to avoid the most severe impacts is fast moving out of reach/Fortune Magazine. See https://fortune.com/2023/03/20/worlds-top-climate-scientists-issue-blunt-warning-our-chance-to-avoid-the-most-severe-impacts-is-fast-moving-out-of-reach/ In September 2023, UN Secretary General Antonio Guterres called for G20 leaders to reshape financial rules after the latest UN Report called for decisive action. See https://unfccc.int/sites/default/files/resource/sb2023_09_adv.pdf
14. Jessel, E. 2023. Calls for more urgency in climate-proofing critical infrastructure. New Civil Engineer. See https://www.newcivilengineer.com/latest/calls-for-more-urgency-in-climate-proofing-critical-infrastructure-06-03-2023/?eea=MVc1YXZibDFPSmRkZ0JoQWk4cW8xc1g1ZkM4TCtWWXI3b3pCVURKY09mQT0%3D&utm_source=acs&utm_medium=email&utm_campaign=CONE_NCE_EDI_ALL_MORNING_060323&deliveryName=DM121985
15. Wolf, M. 2019. FT The US debate on climate change is heating up. https://www.ft.com/content/2a0d4caa-337c-11e9-bb0c-42459962a812
16. O'Brien, R. and M. Rees. 2023. You've got to release the energy and untapped abilities that are in the cities. Marvin Rees talks to Rachel O'Brien. RSA Journal. 169(1(5592)): 20–25. https://www.jstor.org/stable/48719748..
17. Doerr, J. 2021. Speed&Scale: A global action plan for solving our climate crisis. Penguin Business. UK. See https://speedandscale.com/?gclid=Cj0KCQiAqbyNBhC2ARIsALDwAsAuCOEHO3q08FTiA0fAg8K38-_dYDgqEfpL06Omi9DxGb5RrDcRCacaAh3hEALw_wcB
18. Nersisyan, Y. and L.R. Wray. 2020. How to Pay for the Green New Deal. Levy Economics Institute of Bard College Working Paper No. 931. See https://www.levyinstitute.org/pubs/ppb_148.pdf. Murphy, R. and Hines, C. 2010. Green quantitative easing at http://www.financeforthefuture.com/GreenQuEasing.pdf and Green New Deal Group at https://greennewdealgroup.org/. See also Advancing a Green New Deal at https://www.globalgreennewdeal.org/about
19. Songwe, V., N. Stern and A. Bhattacharya. 2022. Finance for climate action: scaling up investment for climate and development. LSE. Grantham Res Inst Climate Change and the Environment. See https://www.lse.ac.uk/granthaminstitute/publication/finance-for-climate-action-scaling-up-investment-for-climate-and-development/?utm_source=substack&utm_medium=email
20. HM Government. 2023. Powering up Britain. See https://assets.publishing.service.gov.uk/government/uploads/system/uploads/attachment_data/file/1147340/powering-up-britain-joint-overview.pdf HM Government. 2023. Mobilising Green Investment. See https://assets.publishing.service.gov.uk/government/uploads/system/uploads/attachment_data/file/1147377/mobilising-green-investment-2023-green-finance-

strategy.pdf and HM Government (2023) 2030 Strategic Framework for International Climate and Nature Action. See https://assets.publishing.service.gov.uk/government/uploads/system/uploads/attachment_data/file/1147393/2030-strategic-framework-for-international-climate-and-nature-action.pdf and Bank of England (2023) Bank of England report on climate-related risks and the regulatory capital frameworks. See https://www.bankofengland.co.uk/prudential-regulation/publication/2023/report-on-climate-related-risks-and-the-regulatory-capital-frameworks

21. Feffer, J. 2023. Russian Green Deal: Light at the End of the Tunnel Nation of Change https://www.nationofchange.org/2021/12/23/russian-green-deal-light-at-the-end-of-the-tunnel/
22. China Global Television Network (CGTN). 2020. Full text: Xi Jinping's speech at the General Debate of the 75th Session of the United Nations General Assembly. https://news.cgtn.com/news/2020-09-23/Full-text-Xi-Jinping-s-speech-at-General-Debate-of-UNGA-U07X2dn8Ag/index.html
23. BBC Global Survey: A World Divided? https://www.ipsos.com/en/bbc-global-survey-world-divided
24. Wolf, M. 2023. The Crisis of Democratic Capitalism. Penguin Books. London. https://www.penguin.co.uk/books/305263/the-crisis-of-democratic-capitalism-by-wolf-martin/9780241303412
25. Fragmented groups such as Da'esh, and Al-Qaida are exploiting instability, human suffering, and conflict in Africa.
26. Jeffreys, B. 2022. Children's mental health: huge rise in severe cases. BBC analysis reveals. BBC News. UK. See https://www.bbc.com/news/education-60197150
27. Ratanjee, V. and Robison, J. 2022. Trust Is in Decline: Here's How to Rebuild It. Gallup Workplace. See https://www.gallup.com/workplace/393401/trust-decline-rebuild.aspx Americans' Confidence in Major U.S. Institutions Dips (gallup.com)
28. Climate Change Committee. 2022. Progress Report to Parliament. https://www.theccc.org.uk/publication/2022-progress-report-to-parliament/ See also the Climate Change Committee progress report of 2023 which states baldly that '*The second National Adaptation Programme has not adequately prepared the UK for climate change. Our assessment has found very limited evidence of the implementation of adaptation at the scale needed to fully prepare for climate risks facing the UK across cities, communities, infrastructure, economy, and ecosystems'. 'The next National Adaptation Programme must make a step change.*' See https://www.theccc.org.uk/publication/progress-in-adapting-to-climate-change-2023-report-to-parliament/#downloads
29. Gayle, D. 2022. Court orders UK government to explain how net zero policies will reach targets. Guardian. UK. See https://www.theguardian.com/environment/2022/jul/18/court-orders-uk-government-to-explain-how-net-zero-policies-will-reach-targets
30. United Nations Climate Change. 2022. Even after COP27 (Conference of the Parties). See https://unfccc.int/event/cop-27) in November 2022, progress is inadequate. Admittedly a major milestone was achieved—agreement on a fund to help deal with loss and damage in developing countries, something they have been seeking for nearly three decades. But there was no agreement on where the funds will come from. In 2015 in Paris two temperature goals were agreed. To keep the rise well below 2°C above pre-industrial levels, and to attempt to keep the increase to 1.5°C. We know that 2°C is not adequate, so COP26 in Glasgow agreed to focus on a 1.5°C limit and to return each year to strengthen the much-needed measures in a process called the ratchet. At COP27, some countries tried to abolish the ratchet. They failed, but a resolution to have emissions peak by 2025 was taken out. A provision was made to boost low-

emissions energy but that could mean many things—even including gas, which has lower emissions than coal. In Glasgow, a commitment was made to phase down the use of coal. At COP27, some countries wanted to include a commitment to phase down all fossil fuels. After intense wrangling, their proposal failed. Urgent changes to the World Bank were called for since it has not provided enough funding to help poor countries. The Intergovernmental Panel on Climate Change (IPCC), set up by the UN, warns of catastrophic impacts that can only be averted by sharp and urgent cuts in greenhouse gas emissions. COP27 refers to tipping points such that the climate does not warm linearly, and we risk tripping feedback loops and rapid escalations. Medical professionals are seeing a clear link between global warming and health. Globally we need no more denying. Extreme weather is coming at us at an alarming pace. People are already dying through frequent heatwaves, floods, wildfires, and droughts. In the UK, the important Autumn Statement 2022 by the then new Chancellor of the Exchequer Jeremy Hunt reaffirmed investment in major infrastructure projects such as High-Speed Rail but he mentioned climate change only once. As of now we are marching ahead along the path that leads to the killing of the golden eggs of the golden goose.

31. Vass, S. 2020. Modern monetary theory: the rise of economists who say huge government debt is not a problem. The Conversation, Modern monetary theory: the rise of economists who say huge government debt is not a problem (theconversation.com).
32. Wikipedia https://en.wikipedia.org/wiki/Golden_Rule Moral relativism is metaethics – the idea that truth or justification of moral judgments is not absolute, but relative to the moral standard of some person or group of persons and controversial. See https://plato.stanford.edu/entries/moral-relativism/
33. Common humanity refers to people being social animals that cooperate. It is the basis for civil, political, economic, social, and cultural human rights. Nelson Mandela is quoted as saying '*To deny people their human rights is to challenge their very humanity*'. [https://www.goodgoodgood.co/articles/humanity-quotes]. Eleanor Roosevelt gave a speech at the United Nations called '*Where Do Human Rights Begin?*'. This is part of her speech. '*Where, after all, do universal human rights begin? In small places, close to home—so close and so small that they cannot be seen on any maps of the world. Yet they are the world of the individual person; the neighbourhood he lives in; the school or college he attends; the factory, farm, or office where he works. Such are the places where every man, woman, and child seek equal justice, equal opportunity, equal dignity without discrimination. Unless these rights have meaning there, they have little meaning anywhere. Without concerted citizen action to uphold them close to home, we shall look in vain for progress in the larger world*'. For more details on the Universal Declaration of Human Rights. See https://www.amnesty.org.uk/universal-declaration-human-rights-UDHR and https://www.un.org/en/about-us/universal-declaration-of-human-rights
34. Revealed: Trust in politicians at lowest level on record. 2021. The Progressive Policy Think Tank. https://www.ippr.org/news-and-media/press-releases/revealed-trust-in-politicians-at-lowest-level-on-record
35. *Ibid* [Wolf].
36. Building Learning Power. 2023. https://www.buildinglearningpower.com/
37. Quinlan Dougherty, T. 2023. Changing Norms to Change Lives Penn Arts & Sciences. https://www.sas.upenn.edu/~cb36/files/chan.pdf
38. Pearlstein, S. 2018. Moral Capitalism; Why Fairness Won't Make Us Poor. St. Martin's Press. New York.

39. Donaldson, T. 2014. Shared values that are lost in translation. Financial Times. https://www.ft.com/content/bfdfff94-b34c-11e3-b09d-00144feabdc0
40. van Riel, Raphael and Robert Van Gulick, "Scientific Reduction", The Stanford Encyclopaedia of Philosophy (Spring 2019 Edition), Edward N. Zalta (ed.). https://plato.stanford.edu/archives/spr2019/entries/scientific-reduction
41. Stallard, E. 2023. What is the UN High Seas Treaty and why is it needed? BBC New. See https://www.bbc.com/news/science-environment-64839763. As I write this a new agreement has been reached at the UN. After more than a decade of negotiations, the countries of the United Nations have agreed the first ever treaty to protect the world's oceans that lie outside national boundaries. The UN High Seas Treaty places 30% of the world's oceans into protected areas, puts more money into marine conservation and means new rules for mining at sea.

Contents

List of Figures

Ten Learning Points about Economics

No.	Learning Point
1	There are three surreptitious assumptions at the heart of capitalist economics. They are that scarcity is central, economics is a science disregarding the way it is practiced and that the idea of the economic man and utility tends to associate capitalism with rewarding greed.
2	There appear to me to be (at least) three important and wide misunderstandings amongst economically influential non-specialists such as politicians. They are how we justify the case for a capitalist economy, how we understand equilibrium, simplicity, and complexity and finally the role of fiat money and debt and the consequences for fiscal and monetary policy and money supply.
3	Government spending creates the money to pay tax and fund government borrowing.
4	Fiat money is not well understood by many people in power—especially politicians—and this leads to wrong-headed policy decisions.
5	The deficit of Governments with fiat currency (Government spend > tax) puts money into the private sector.
6	The purpose of selling Government Treasury Bonds in fiat currency is not to borrow money from investors rather it is to avoid an overdraft in its account at the Central Bank.
7	National debt need not be a burden to our children and grandchildren.
8	Interest on Treasury Bonds can be paid by the Treasury borrowing from its own Central Bank.
9	We should look at net worth, not profit, to assess financial value of all financial entities whether individuals, companies, or governments.

10 Principled capitalism can emerge out of liberal capitalism if we widen utility as personal satisfaction and the profit motive to include all forms of obligations to each other and to nature. In that way we can become citizens of the world defined by our interdependence and community.

Five Axioms of Systems Thinking

No.	Title
1	A system, discriminated from its surroundings, is a set of models of our reality created by people for reasons which give meaning to and determine the purpose of the system.
2	Systems models are holons, i.e., they are both parts and wholes and hence are layered according to levels of detail and abstraction.
3	Holons are connected to certain other holons with which they exchange messages.
4	Systems models change at varying rates, but none are permanent and invariable. Some changes may be unforeseen. Some changes may be small, but some may be 'revolutionary paradigm shifts' involving new ways of thinking.
5	Complex systems often cannot be 'solved' rather they have to be managed to desirable outcomes.

Part 1

Thinking

Preamble to Part 1

Thinking about thinking is hard. The philosopher John Dewey said, '*We do not learn from experience... we learn from reflecting on experience*'. Metacognition is about planning, monitoring, and evaluating one's own experiences and thoughts. Few of us do it but, as we head for possible climate change disaster, it is important to reflect. The Buddha understood this. One translation of Dhammapada Verse 2[1] is '*All that man experiences springs out of his thoughts. If his thoughts are good, the words and deeds will also be good. The result of good thoughts, words and deeds will be happiness. This happiness never leaves the person whose thoughts are good. Happiness will always follow him like his shadow that never leaves him*'. In other words, think wrong and suffer. Think right and be happy.

In Part 1 we will think about ourselves, our natural world, our systems (social and physical) and two of the most important factors in climate change, engineering (including technology), and economics.

The central messages of Part 1 are first, that our thoughts can change the way we behave. Collectively we have become rather self-centred and greedy. One of the major reasons is that we have come to feel that to be happy we need money. Having enough money (however defined) is necessary but not sufficient. Quickly that desire for money can morph into greed for more and more. We come to think that money shields us from anxiety, and loneliness. We think it gives us a sense of security and financial stability. But these attitudes derive largely through the way our capitalist economy is run with the overriding motive of profit for shareholders. We need a different form of capitalism—principled capitalism–driven by moral virtue.

Second, we cannot have personal wellbeing without nature's wellbeing. We have made progress in understanding we are part of nature. 'Nature Connectedness' is about our emotional connections and responses to nature which help regulate our own feelings and keeps us mentally healthy[2]. We are rebooting our policies and practices for the natural world[3] but we need to speed up progress[4].

Third, we need to rethink how we manage our systems both physical (such as a transport infrastructure—bridges, railways, roads, communications, etc.) and human/social (such as health and social services—integrating them or joining them up into a smoothly run up-to-date service). The concept of a system is somewhat abstract—but important and worth getting your head around. I define a system as something, discriminated from its surroundings, as a set of models of our reality. A system is created by people for reasons which give meaning to and determine its purpose. We look at the nature of models in Chapter 2. Systems are all around us. The reasons why we identify them maybe because we are curious and want to understand (science) or because we want to modify the world in some way to improve the human condition (engineering, medicine, politics, and economics) or express our emotions (art, religion). Indeed, our purpose in identifying a system is our highest goal. That goal provides us with meaning and motivates us to put in effort to add value. Identifying purpose draws on our emotional intelligence to help us reflect on and understand questions why we do things. In practice many of our important systems do not work well because they are not 'joined-up'. People tend to work in silos with poor communication between functional departments and professional disciplines. To tackle this issue, we need to go beyond scientific reductionism to the 'bigger picture' of systems thinking as we shall explore in Chapters 2 and 7.

Fourth, there is a need for a better understanding of the relationship between science, engineering, and technology. The new technologies, such as artificial intelligence, will be and are already being already, as disruptive as the steam engine was in the 19th century. Technology is a subset of the profession of engineering. Engineering is more than technology—it requires engineers with practical wisdom. Practical wisdom derives from Aristotle's *phronesis* that has been lost in the mists of history. The word engineer derives from the Latin *ingeniator* or someone who is ingenious. The future requires that capacity of ingenuity and innovation.

Fifth, and arguably the most important, is a rethinking of economics. Much of the book is devoted to this because there are deep divisions and misunderstandings about economic policy and few non-economists understand what is being done in their name. The most important problem is that widespread fear of government deficits and debt are holding back much needed investment, for renewable energy, R & D, education, and training and reducing inequalities of wealth. Governments with fiat currency must invest in specific targets whilst avoiding inflation.

End Notes

1. Ven. Weagoda Sarada Maha Thero. 2021. Dhammapada (Illustrated). Wisdom Library. See https://www.wisdomlib.org/buddhism/book/dhammapada-illustrated/d/doc1084242.html
2. Miles 2020. Finding Nature. A new relationship with nature: what it means and what we can do. University of Derby. See https://findingnature.org.uk/2020/04/08/a-new-relationship-with-nature/
3. UK Environmental Act 2021 provides the UK Government with powers to set new binding targets for air quality, water, biodiversity, and waste reduction. These targets should be ambitious, meaningful, and informed by experts. The Act also established a new environmental watchdog, the UK Office for Environmental Protection (OEP), which will hold the Government and other public bodies to account and ensure that environmental laws are complied with. See https://www.legislation.gov.uk/ukpga/2021/30/contents/enacted
4. ClientEarth uses the law to fight climate change, tackle pollution, defend wildlife and protect people and planet. They say '*...a future in which people and planet thrive together isn't just possible: it's essential*'. See https://www.clientearth.org/about/. Regarding the Act they say, 'The Act that is now law falls short of the world-leading ambition that the government initially set out to ensure'. See https://www.clientearth.org/latest/latest-updates/news/why-the-uk-environment-bill-matters/

Chapter 1

Be Careful What You Think

Because Your Thoughts Run Your Life

(Proverbs 4:23: ICB version)

AN AESOP FABLE

The pigeons were being persecuted by a Kite who, every now and then, swooped down and carried off one of their number. So, they invited a Hawk into the dovecote to defend them. But they soon repented of their folly: for the Hawk killed more of them in a day than the Kite had done in a year.

PARABLE OF THE SADHU[1]

A group of hikers were climbing across the mountains of Nepal towards a holy village. They came across a Sadhu, one of the holy Nepalese men who wander the countryside asking for food. The Sadhu was wearing only a few pieces of clothing and shivering from the cold. They gave him warm clothes, food, and drink and one said that they should carry the Sadhu to the nearest village. No one offered to help giving a different reason for not stopping their hike. They left the Sadhu lying on a rock in the sun and continued their journey and no one knows if the man survived.[1]

Man: *I know the advantage of having a poor memory"*

Woman: *"What's that?"*

Man: *"You can enjoy the same joke twice"*

LIFE ON EARTH IS ENDANGERED

Life on earth is the golden egg begat by the goose of planet earth. The goose is exhausted and may soon stop laying eggs and may even die. Life on earth will be no more unless we nurse the goose back into better health.

The Golden Goose

WE KILLED NAURU'S GOLDEN GOOSE

Nauru is a small island in the Central Pacific. It is the third-smallest country in the world behind Vatican City and Monaco. The first European settlers in the late 19th century aptly named it Pleasant Island—beautiful, green, full of life and a haven for migratory birds. Unbeknown to its inhabitants it had its own golden goose. Over thousands of years the accumulated excrement of seabirds and bats created guano—a rich composition of nutrients for farming—a source of phosphorus valuable to the outside world. From 1907, a succession of multinational companies has stripped Nauru bare. It is now a giant opencast big black hole. There are now few signs of life—no trees, no animals, and few areas where people can live. Only twelve miles around the island's perimeter are habitable. The few remaining residents are left with a barren island that cannot sustain its own food supply.

The killing of Nauru's golden goose is one of power, exploitation, and greed. The selling of the entire future of a country for short term gain. After independence in 1968 Nauru had the highest GDP per capita in the world—now it is dependent on foreign aid. It is a story of putting all 'eggs into one basket'. The economy became totally dependent on one commodity and failed to use its short-term wealth to diversify. Nauru is now a detention centre for the Australian government. There is little farmland and severe air and water pollution. Food is imported cheap and heavily processed resulting in increased levels of obesity and ill health. Nauru is an unfortunate example of having what Richard Auty calls a 'resource curse'[2]. Where some countries with abundant natural resources kill their golden goose. They behave rather like some national lottery winners who don't cope well with their newfound wealth.

The story of Nauru is a warning of the damage we can do to our world if we don't change our business models. We have to learn how not to treat our own golden goose—nature itself.

THINKING ABOUT OURSELVES—WE BECOME WHAT WE THINK

When I was 17 my school head teacher Dr Charles told us this story as part of a 'general knowledge' class. The school science labs were being refurbished and there was some scaffolding. One day, the science teacher *'old Grumpy'* (as we used to call him), was demonstrating a science experiment to a class of 13-year-olds. He was showing them how potassium metal reacts in water. You may know that potassium in water melts, and floats, moving quickly on the surface. Hydrogen ignites and the metal is set on fire, with sparks and a lilac flame. There is sometimes a small explosion at the end. One of the children got a bit too close. Old Grumpy, despite his nickname, was actually a good conscientious teacher and he was somewhat concerned. He prudently took the boy to the local hospital A&E just to be sure. There was no problem, and the child was fine. However unbeknown to the school, someone from the hospital rang the local newspaper to tell them what had happened. The headline that night was *'Boy injured in science experiment that went wrong'*. Then the very next morning a small item on an inside page of a national newspaper reported the incident. The next day Dr Charles had a visitor from out of town. As he showed off the school to his visitor on their way to lunch, they passed by the scaffolding. Dr Charles apologized for the mess. The visitor replied, *'Yes I heard you had an explosion recently'*.

Dr Charles' story was the first time that I had come across a newspaper misreporting and exaggerating a minor event. The journalists didn't check any facts—they just assumed that what they were told was correct. Then they exaggerated it very slightly to make a headline. The visitor had seen the reports and he interpreted them in the context of seeing the scaffold. A minor event had erupted into a massive misunderstanding. Dr Charles had to issue a public statement to worried parents and residents. This was not good journalism.

Of course, we are all different and we create and live in our own realities. We pursue our own needs. But making assumptions, without thought or dependable evidence, can mislead and cause harm to others. If we decide on something without any solid evidence, even if we have our own reasons to do so, we should be aware of a potential for hurt. Sometimes we may simply be mischievous but sometimes we may do it for monetary gain (as I suspect the hospital informant did). Often, we make assumptions about other people and events. We may be suspicious, for example, of a partner's fidelity for no reason. The lack of trust that engenders can be pernicious. On occasion, we don't find it easy to identify, examine and change our beliefs. '*We've always done it this way—don't mess up something that works*' is the comeback. When we think we know something we tend to act without much thought about what we might be assuming. But making decisions without thinking things through can be dangerous. Unanticipated things happen. Life can get complicated—especially when we tell a lie—whether a justifiable white lie to save someone's feelings or a downright lie to get what we want.

Joining a group of likeminded people can be incredibly positive. We are social animals and others can help us to enjoy life and do the things we like and are reasonably good at—whether following our religious faith, playing sport, listening to, or playing music, having a good conversation and socializing in various ways. Being part of a group gives us a sense of purpose and builds our own feelings of self-worth. Groups help us to handle the self-doubt that we all suffer from—despite the denials of some.

SOCIAL MEDIA IS CHANGING SOCIAL NORMS

Social media now plays an important role in the mental health of users—especially the young. We are all influenced by the behaviors of others, sometimes positively and sometimes negatively. Some tend to copy their heroes. They may realise that some norms, like fashions, are arbitrary. Social media can promote ethical, responsible behavior and personal growth. But it can also discourage rationality, encourage bad habits, and increase risks to individuals. It has become commonplace to share private information with other users and so, by extension, publicly with third parties, service providers, and institutions. Some US and UK agencies have been told to delete TikTok from their mobile phones because governments are concerned that sensitive data may be sent to China. Complex messages are often reduced to slogans which can increase partisanship. There is a growing body of work that shows that concentration disorders are increasingly common. At the extreme, some social media groups can make people feel worthless and depressed. Increased mental illness can become so severe that people take their own lives. Criminal gangs exploit the vulnerabilities of children to peddle drugs or become involved in sex crimes.

OUR PERSONAL ASSUMPTIONS HELP US TO LIVE IN HOPE AND FULFILLMENT

Some people seem to handle their self-doubt by retreating into themselves and become so introverted, shy, and retiring that it is difficult for others close to them to work out what they are thinking. Some become so extrovert that they seem to be full of their own sense of self-importance. They behave arrogantly, are conceited, and sometimes pretentious. On the flip side I am often impressed by the self-assurance of public figures as they cope with the pressures put on them by regular public media appearances. But out of the spotlight, even they may some suffer from self-doubt and anxiety from the worry of having to be successful. Perhaps they try to rid themselves of depressive thoughts by entertaining others —a particular trait amongst some comedians such as Tony Hancock who took his own life in 1968 after his earlier great successes had taken a dip.

Assumptions, for example, that our partner will be faithful, are part of who we are, and they help to define us and our feelings of self-worth. Some assumptions we can control—such as our level of honesty, our integrity, and our faith in whatever we choose to have faith in whether a god, humanity in general, our friends and relations. Some assumptions allow us to influence things—such as the behavior of a friend. But there are things over which we have no control no matter what we assume—such as the weather. Other factors that make us who we are, include where we were born, where we live, who looks after us when we are small, the friends we choose, who inspires us, who teaches us, the death of loved ones, experiences, and accidents of all kinds.

Ultimately no-one else can choose our thoughts. But we are influenced, as I have said, by social norms. Cristina Bicchieri has studied the effect of social norms on how we think and behave[3]. Much seems to depend on people's expectations. As we grow up, we are affected by our parents, teachers and the people and events around us as we mature. We tend to do what others do—the social norm. We don't want the disapproval of others. But our thoughts as we develop are ours and ours alone. Some people call this our mindset. Others call it our worldview and philosophers refer to the German word for it—*weltanschauung* after the famous philosopher Immanuel Kant[4]. To change our worldview, we need to be convinced there is good reason. There can't be a bigger reason than the damage and destruction of climate change. But we need leaders and trendsetters to help us and show us the way.

SOCIAL NORMS—GOOD AND BAD

When we think some thoughts repeatedly and they are reinforced by like-minded people, we may begin to believe they are true—even without vindication. We may join political parties, factions, sects, and other groups. Joining likeminded groups can be good. They are of immense benefit to professionals by sharing experiences, information, and knowledge. Unfortunately, some groups are bad—malfeasant, dishonest and corrupt.

As these groups, institutions, societies, syndicates, and gangs grow and influence, they start to set standards of who can join. The good ones create the professional standards to qualify as, say, a doctor or lawyer. They decide what tests people must pass to join. They define the ongoing worldview and social norms of the group. The bad, such as criminal gangs may require new recruits to prove themselves by committing a specific 'initiation' crime.

To reduce and eliminate corruption we have to change social norms into ones that strive to reinforce the good and reduce the bad. That means creating a greater awareness of moral disengagement[5]. Why do some people act inhumanely or corruptly without being stressed? Most people do not behave unethically, except when they can justify it to themselves. They may think they are acting in a manner that has moral value. For example, taking a bribe to benefit the company or as a 'lubricating oil for development' of a project. Albert Bandura suggested eight mechanisms for moral disengagement[6]. The first is moral justification or reconstructing the affected behavior—for example, when a soccer player breaks the rules to score a goal. Second is euphemistic labelling—such as talk of 'bending the rules.' Third is making an advantageous comparison with more harmful behavior—such as comparing abusive language in sport by saying it's not as bad as physical injury. Fourth is displacement of responsibility through pressure from the boss or because others are doing it. In other words, blaming someone else for making them act in a corrupt way. Fifth is diffusion of responsibility—for example, by acting as a group and diluting individual responsibility in the group. Sixth is distortion of consequences by minimising or ignoring or disbelieving harm—for example by saying the harm was not as bad as it appeared to be. Seventh is by dehumanizing the victim—for example, calling him a dog or savage and hence below contempt in some way. Finally comes the attribution of blame—for example, by saying the other party acted first.

It seems, from the little research that has been done, that social norms strongly affect individual behavior. If the peer group is critical of self-justification, then moral disengagement is reduced. The behavior of the individual becomes transparent and generates shame.

People with a proneness to guilt and shame, anticipate and see consequences. They reflect on their internal morals. They anticipate and avoid bad moral emotions and unethical behavior.

PROFESSIONAL ASSUMPTIONS SET STANDARDS OF COMPETENCE BUT CAN GENERATE SILOS

Professionals join expert groups to qualify, learn and share. For example, engineers join specialized separate institutions of civil, mechanical, electrical engineers (and more) by passing exams. Economists may qualify as public or certified accountants or chartered economists. These groups or societies develop their own ways of understanding the world—they develop a shared mindset. Importantly, professional institutions provide the public with assurance. By setting standards of competence, they protect us from error and even mischief. We know if we ask one of them to do something for us, we can be assured they are not some cowboys pretending to be something they aren't. For example, when we go to see a doctor, we know from their membership of the Royal College of General Practitioners that they are competent to treat us—they have passed the exams.

Unfortunately, as valuable as these groups are, they can become isolated 'silos' of thought—with little cross fertilization with other groups. Resulting inefficiencies can cause real harm in some cases when communications are inadequate. They may even result in negligence leading to death in catastrophes such as the fire at Grenfell Tower in London in 2017.

Economists also divide themselves into schools of thought according to their interpretations of economic theory. Early thinkers are called classical economists. Then came neoclassical, Marxian, Keynesians, monetarists, and many more including labour, environmental, financial, econometrics, international, health and public economists. They all base their understanding of economics on different assumptions although there are many held in common. Modern economists refer to a deep division between the traditional orthodox and modern heterodox approaches. We will explore these divisions later.

I think it is helpful to see the formal assumptions that professionals make, in the context of the kinds of assumptions we make in everyday life. By more formal I mean methodical, precise, and conventional. For example, some of our daily assumptions are so regular that we hardly notice them, and we take them as factual. Every one of us assumes that the sun will rise and set every day. We've never known it not to and we have no reason to doubt it. But there have been times in history when the sun has been blocked out. In 536 AD Europe, the Middle East, and parts of Asia were plunged into 18 months of solid darkness by a fog from a gigantic volcanic eruption in Iceland. According to Michael McCormick, a Harvard University archaeologist and medieval historian[7], the world did not show signs of recovery until 640 AD. *'It was the beginning of one of the worst periods to be alive, if not the worst year'*, said Dr McCormick. The fog caused a darkness over the northern hemisphere for 18 months with severe effects on the climate. Temperatures in the summer of 536 fell between 1.5°C and 2.5°C. A more recent, but less serious, volcanic shock came in June 2010 when a series of eruptions at Eyjafjallajökull in Iceland caused massive disruption to air travel across Western Europe.

Our confidence in even the most regular events on earth cannot be total. We can and do assume them for everyday life. We take them for granted—but we can't be certain of them logically. The word logical is important here. We will come to it later and in Chapter 3 but briefly for now by logical we mean using the principles of logic for making formal dependable inferences. Arriving at a conclusion from clearly stated assumptions.

In the past and present most of us have taken our planet for granted. We can no longer afford to do so. Our misplaced confidence has relied on the laws of physics to assume and explain many physical phenomena. They have held good in the past so it seems reasonable to assume that they will continue to hold good in the future. But no matter how many times we see one thing as causing another we cannot logically be certain that will always be the case. Put formally those laws are not logically entailed by the observed instances no matter how numerous. Philosophers call this the problem of induction—inferring the general case from a particular one. Bryan Magee summed it up nicely when he wrote '*Even when all past futures have resembled past pasts it does not follow that all future futures will resemble future pasts*'[8].

Later in this chapter and in Chapter 2 we are going to look at in more detail some of the assumptions made by economists and engineers. These kinds of assumptions made by professionals are more formal than our everyday ones. The reason is that professionals must be able to justify their decisions to us, the public, and through us to the law of the land. They are accountable for what they do. They have what lawyers call a public 'duty of care'. That means they have obligations to act towards others adhering to a standard of reasonable care in accordance with the standards claimed by their profession. To fulfil those obligations all ideas must be shared and discussed at length. It is a kind of formalization of a social contract between a professional and those who may be affected by his/her decisions. So, it is important to try to tease out why any assumptions, whether individual or collective, that might hinder progress towards our goals.

Another way of seeing it is that professional assumptions need to be testable. The philosopher Sir Karl Popper distinguished scientific knowledge from the rest by requiring it to be testable. By that he meant testable by physical experiment or if that is not feasible then testable through critical discussion. We shall explore this important distinction in more detail later because it is significant in deciding between news and fake news, between fact and fiction, between things we can depend on and things we cannot.

One way of thinking about our assumptions, thoughts, and beliefs, whether personal or professional, is that they become a cognitive 'pair of spectacles' through which we see the world and everything in it. We interpret our world and our part in it through those lenses. They become a filter through which we interpret evidence that matches what we believe to be the case, i.e., true. In professionals this can and does[9], as we said earlier, lead to silos of separated and fragmented thought which are deep and highly specialized, but which do not relate to each other at all well. It can result in organisations where resources are not shared across departments. It can make 'joined-up' integrated working together in teams more difficult. Ideas are not exchanged, production slows down, inefficiencies increase. Employees and customers feel frustrated growth of business prevented.

As noted, earlier Buddha said, '*We are what we think ... with our thoughts we make the world*'. If the Buddha was alive today, I am sure he would add—the way we are thinking, and behaving must change. If life on Earth is to survive massive harm or even possible extinction, we must think and do things differently.

In businesses and in everyday life we need to collaborate better. The ways to do this are well known but often ignored. Transparency of news, information and best practices is one of the first requirements. Others include learning together to develop our knowledge and skills, building trust through clarity of democratic values and obligations to each other and our environment, communicating shared experiences and purposeful goals, and devising strategies to reach them. In short getting everyone, who has a stake, engaged in the way we live. We'll see later these are all aspects of what Jon Alexander[11] and others calls the 'citizen story'. They want us to move from the 'consumer story' to the citizen story—where we all see ourselves as citizens of the world as well as our own communities and nations. The first stage of that journey is to become critical of our own assumptions. To keep our minds open to alternative viewpoints.

History tells us that in times of war a common enemy creates a shared common purpose. We have that war, and the enemy is our changing climate. To fight that war, we will have to come together and that will almost certainly mean changing some of our individual and collective assumptions.

BE CRITICAL OF YOUR OWN ASSUMPTIONS

Many years ago, I had a friend called Jim who always knew what he thought about everything—and he proclaimed it loudly to all and sundry. He was a nice enough fellow but to put it bluntly he had a fixed mindset. I never detected him changing his mind even when others proved him wrong. It was if his self-worth depended on the strength and emphatic expression of his views. Claire, another friend, on the other hand was a distinct opposite—she changed her mind like the weather (in the UK) and every opinion seem to derive from the last person she spoke to.

Whether you are a Jim or a Claire, or most likely somewhere in between, then you may or may not appreciate that to improve and adapt we must first become aware. To change we must become critical of our thoughts and assumptions. For example, a lot of people struggle with negative emotions. To overcome them we must first identify them. Only then does it become possible to explore the reasons for them. We have to try to learn to replace them with much more positive ones. If we can become aware of our assumptions, then we are able to test them out. If they turn out to be inadequate in any way, then we can try to change them—not an easy thing to do in any circumstance.

The idea of being critical of our assumptions goes very deep—even into the heart of knowledge itself. It underpins that most secure form of human knowledge—science. Scientists are constantly examining their assumptions as carefully as they can for the situation they are addressing. If we can, we should all, scientists or not, try to test out our assumptions in some practical way. As I said earlier that is how the philosopher Karl Popper distinguished science from non-science. But some assumptions are not testable—in that case he said then we should, at least require them to be logically consistent. So, what are our most basic assumptions?

NATURAL LAW IS A GOOD STARTING POINT

The mathematical biologist Robert Rosen[10] defined what he called *Natural Law* with two assumptions. Both are so fundamental that it is difficult to see any others replacing them. Rosen's first assumption is that the succession of events or phenomena that we perceive in the world, and in some specific context, are not entirely arbitrary or whimsical. There are relationships that are manifest. Rosen's second assumption is that these relations are at least in part, capable of being perceived and grasped by the human mind.

These deceivingly simple statements led to dense interpretations by him that we'll need to unpack. He said that we can compare structures of 'logical entailment' in formal systems with structures of 'causal entailment' in natural systems. Put simply logical entailment just means 'to be a necessary consequence of' or in other words to follow logically. Causal entailment is practical. Causal entailment just means that an 'effect is a necessary consequence of a cause'. If I drop a wine glass on a hard stone floor (cause) it shatters (effect)—but not always—occasionally the glass may survive intact. Causal entailment is difficult because it depends on circumstances or context. It is variously uncertain.

A natural system is the reality of our existence—the natural world—including us. A wine glass breaking occurs in a natural system. A formal system, on the other hand is a theoretical one which is logically consistent. It is an 'axiomatic system'. That means it is an abstract structure of assumptions called axioms with rules based on a logical calculus. One everyday example is arithmetic. One of the eight axioms of arithmetic, for example, is that $x + y = y + x$ or $1 + 3 = 3 + 1$. The order in which you add up numbers doesn't matter—you will still get the same total (unless you make an error). More generally a formal system is mathematics and theories of formal logic with axioms based on set theory. Mathematics is a formal language with stated axioms from which theorems can be derived by deductions. In other words, using the axioms and rules one can deduce or build theorems which follow logically and hence are true within the system. Unfortunately, in the 1930s, a mathematician called Kurt Gödel showed that formal systems are incomplete. It was another complexity we could do without—it makes theoretical matters more difficult[11]. This leads us to another dense interpretation that we need to unpack. Rosen's Natural Law effectively assumes an *a priori* synthetic relation between logical reasoning and causality that depends on context. Here the Latin phrase *a priori* simply means before the fact. Synthetic means that a denial would be a logical contradiction—for example, a wine glass is made of glass. Context describes the circumstances surrounding a situation. For example, Newton's Laws of gravitation assume that nothing travels at a significant proportion of the speed of light.

We could sum all of this up by saying that Natural Law assumes that nature (including we humans) is orderly, and describable but not evident or obvious. It is simple common sense, yet at the same time, complex. This is a theme that we will return to many times—nature is at one both simple and complex.

THINKING ABOUT NATURE

Our wellbeing relies on the wellbeing of nature. Our relationship with nature is crucial. According to Friends of the Earth seven out of ten people in the UK admit they are losing touch with nature. A third admit they could not teach their children about British wildlife[12]. Many of us have, in the past, and many still do, take the natural world for granted. Common species like bees, hedgehogs, starlings, and house sparrows are in trouble—going missing from our gardens, streets, and neighbourhoods. If the bees and birds lose out big time, then so do we. We must bring nature back from the brink because it's in our own interests to do so. We seem to be the only species on Earth that actively destroys its own life-support systems. For example, few of our rivers in the UK are unpolluted enough to swim in and wildlife is threatened. We can't let nature continue to decline we must act to repair, restore, and maintain it. We must invest to clean up our rivers. Sir David Attenborough is quoted as saying "*The effect that human beings are having on the natural world is profound. Because we are out of touch with the natural world... most of us don't see the effects*".

We should educate our citizens to reengage. Friends of the Earth write '*For children and adults alike, daily contact with nature is linked to better health, less stress, better mood, reduced obesity—an amazing list of features no other product can ever match*'.

How we perceive and know nature, and the language we use to describe it has implications for conservation policy. According to Hatty et al.[13] people seem to use three categories: descriptive (e.g., plants, animals, landscapes), normative (e.g., conservation, balance, life), and experiential (e.g., activities in nature, positive emotions, aesthetic qualities). They found that people who described nature in experiential terms were more likely to have participated in environmental volunteering, citizen science, picking up litter, and community gardening in the past year than those who used descriptive terms. Researchers at the University of Derby[13] found that we should focus on five types of activity: tuning our senses, responding with our emotions, appreciating beauty, celebrating meaning, and activating our compassion for nature. They call it 'nature connectedness'. They found that sensory contact, emotion, beauty, meaning, and compassion connected people to nature more effectively than the traditional approach of facts, figures, and science. Positive emotions make us feel good, have pleasant experiences, and fulfilled desires. Our ability to function well psychologically is often referred to as '*eudaimonic wellbeing*' or reaching our potential. It includes autonomy, self-acceptance, meaning and purpose in life, and personal growth. Eudemonic wellbeing is related to hedonic—the pursuit of pleasure—wellbeing through different motives, behaviors, and experiences. People with high levels of both types of wellbeing are said to be flourishing.

LIFE IS BOTH SIMPLE AND COMPLEX

We have said that life is simple and complex at the same time. When we are children we take things, more or less, as they come. We don't necessarily see the complexity that is there in front of us. For example, when I first started to play football (soccer in the USA) for my school team, I did not see the difference between football results and arithmetic. I knew that if a number, say x, is bigger than another number, say y, and y is bigger than z then it follows logically that x is bigger than z. But when my team, Hardwick School beat Allenton, and Allenton beat Normanton then I thought we, Hardwick should easily beat Normanton. How wrong I was. We did win but it was a tough game settled by a last-minute goal. Now, after a formal education in engineering, I know that, technically, arithmetic is transitive, and football isn't.

Likewise, I didn't realise how fragile childhood friendships could be—small hurts and resentments led to major arguments and fights. Children are rarely equipped to discuss conflicts—especially with their parents. They want the attention and can't understand the problems others may have. They may resent a father who they think is too busy to pay attention to them. From his point of view, he is doing his best in a complex world to keep them fed and secure. They find it difficult to forgive and forget any neglect—as indeed do many adults. Another good example of complexity masquerading as simplicity was at Sunday School at my local church. My parents sent me there when I was a small boy. I learned to accept, without questioning, the idea that when you die and if you have been good, then you go up to heaven. If you were bad, then you went to down hell. Heaven was somewhere in the sky and hell somewhere down below. A simple solution to a complex question that many children accept—especially if someone in the close family dies and the child asks, '*What happens to Grandad now?*' Only in my late teens did I begin to wonder, think about, and question the complexity behind these kinds of simple answers.

One of the major scientific achievements of the 21st century is that collectively we have become all too aware formally of complexity. To scientists and professionals, the world is a much more complex place than we previously thought. We are now beginning to understand that our previous theories about the way the world works, which seemed to explain everything, were overly simple (though it didn't seem that simple when we were learning it at school). Collectively, scientists were overconfident in their belief that they understood the world we live in. Like children we hadn't seen the complexity before our very eyes. It was there in front of us, but we hadn't seen it. We thought that our explanations of the physical world were the 'truth,' and the world was rather straightforward. There were just a few anomalies to sort out and we would understand the mind of God. Even the eminent physicist Stephen Hawking speculated on us knowing the mind of God. Again, how wrong we all were. I use the quotation marks around truth here to signify absolute truth in all contexts. Now we are realizing that we can only talk in terms of truth in context. Edward de Bono (1933–2021) gave a good example. He asked in a lecture, in the 1980s, when I was in the audience, '*What happens if I let go of this stick in my hand?*' The reply from the hall was '*It falls to the ground!*' '*Yes, you are right*' said de Bono, '*but it also depends on context. If I am on the earth's surface on dry land—yes it falls. But if I am under water it rises. If I am in a space capsule circling the Earth, then it stays where it is—it floats because there is no gravity'.* That's when it dawned on me that context is all important when determining truth.

Modern media and information technology has brought enormous benefits but lots of complex unintended consequences. Before we look at some of them, we should be clear what a complex system looks like. A complex system is one where things are related in ways we didn't expect. Indeed, some say that everything is related to everything else. A flap of the wings of a butterfly in the UK can cause a tornado in Florida[14]. The word cause is central here and we shall come across it many times. It seems as if nature is a large network of components without any form of central control. Some cannot accept that idea and attribute the control to something 'out there' that is greater than any one of us. We all have sense of our inability to understand everything there is to know—indeed some of it is indeed unknowable. Common sense tells us that no-one knows everything—there will always be unknown unknowns—things we can't know that will throw up surprises from time to time. Unintended consequences are surprises. Who would have thought that the development of the contraceptive pill for women in the 1950s would lead to so many one parent families in the 2000s. Who anticipated the invention of the transducer would lead to us seeing pictures from the Moon or sharing videos on mobile phones? Who thought that identifying the DNA molecule would lead to the arrest of murderers?

THINKING ABOUT SYSTEMS

Complexity, seen or unforeseen, is one of the reasons why our previous attempts to understand the world around us were divided into different disciplines—there is so much to study that we divided to conquer. In science the traditional divisions are physics, chemistry, biology, mathematics, chemistry—each of which have subdivisions. In the humanities we have art, languages, etc. But these disciplines all overlap to form the whole which we neglected. Unfortunately, subject silos have formed. Very few people see the overview—how it all fits together. Of course, understanding in one area such as mathematics helps understanding in another such as physics—those two are so closely related that there are scientists that call themselves mathematical physicists. Mathematics is a language for logical deduction and formal reasoning.

General Practitioners (GP) doctors have to take an overview of a patient's condition—but, of course, they immediately refer to a specialist, such as a cardiologist, when they feel patient's problem goes deeper than they can deal with.

As I said earlier the idea of a system is rather abstract. A system is something, discriminated from its surroundings, as a set of models of our reality created by people for reasons which give meaning to and determine the purpose of the system. We'll talk about models in Chapter 2. Systems are all around us. A good way to think about a system is by asking questions based on the 'six serving men'[15] of why, how, who, what, where and when. My first axiom of systems thinking is about questions 'why'. We do we build our models of systems? The answer is for a purpose—a reason. The answer to the why questions is a potential[16] that drives a flow of change. In physics a potential is, for example, gravitational or electromagnetic. The potential is the work or energy transferred per unit of mass that is needed to move an object in what physicists call a field. In other words, potential is a capability to become. A field is a region with a physical quantity such as a force or energy. Fields are basic to quantum physics. In them particles are perturbations in their own fields that interact with others such as the Higgs field that imparts mass. The potential creates a flow of movement. In the field of human interactions, potential is the purpose (why) that drives the flow of change in in people (who), things (what), context (where) and through time (when).

The purpose of science is to understand and hence to know. It is used to predict what the future may be like. It helps us to interpret what may be happening to you as you read this. For example, there are physical forces in and on your body, chemical reactions, and biological processes in your cells. Your skeleton, your immune systems, your circulatory and nervous systems are all working all the time you are alive. Traditionally scientists and medics have studied these systems separately. But the deeper they studied the more specialized they had to become—there is so much detail to absorb and understand.

More recently, scientists have begun to realise that is not enough—they have to see the whole as well as the parts. You, as a whole person, are made up of subsystems that work together. Your sub-systems are, for example, your bones and muscle, nervous, blood circulation and immune system, etc. They interconnect to produce attributes that 'emerge' from the interactions between them such as your ability to walk and talk. But it doesn't stop there because you are a part of your family and community. As the poet John Donne wrote[17], *'No man is an island'* by which he meant that people need each other and are one piece of a greater whole that is humanity. Your community is part of a town or city, region, nation, and also organisations, religious and racial group. Ultimately, as Donne said, we are all human and part of the natural world—global citizens. This hierarchy of parts and wholes is found in all natural systems. They all have emergent attributes—ones that the subsystems do not have separately. When the sub-systems interact with each other these new attributes are created. For example, whilst your football team may be having a successful run, then they have an emergent property of 'being on a good run'. None of players can win on their own but the team can. Success is an emergent property of the football club. This has led many to believe that all systems are both parts and wholes at the same time. Arthur Koestler[18] called them *holons*. A whole new field of systems thinking has developed as a basis for systems science and systems engineering. Although economists refer to economic systems (command, market, etc.) there is little evidence of systems thinking[19] in economics, however.

THINKING ABOUT ENGINEERING AND ECONOMICS

Neither discipline is well understood by non-specialists including many politicians, opinion formers and the electorate. For example, David Clarke of PositiveMoney reported the results of a poll[20] in 2017 that 85% of UK MPs were unaware that new money was created every time a commercial bank extended a loan, while 70% thought that only the government had the power to create new money. Around 70% of UK MPs believed, wrongly, that only the government has the authority to create all new money. Politicians often tell us that there is 'no magic money tree'[21] to pay for vital services but they are worryingly ignorant of where money comes from.

There are many myths about engineering. Many people see engineers as introverted geeks, with no social skills, who love mathematics and detail. They see engineering as difficult and not creative and dominantly male. Whilst there are engineers who fit this profile the vast majority do not. As we will see in Chapter 2 engineering is so badly understood it is often confused with technology and science. For some the typical engineer is somebody who fixes things. There are only a handful of MPs in the UK who have scientific qualifications and even fewer engineers. Only three US Senators are engineers with seventeen sitting in the House of Representatives[22].

Engineering is More than Technology

I wonder what image the word engineer conjures up in your mind. Do you see a mechanic in greasy overalls with a beer in one hand and a spanner in the other? Do you see a woman in a high visibility jacket or a man in a suit wearing a hard hat? Or do you see a hairy boffin in a lab coat with a note board and peering into some complicated equipment like a jet engine? Likewise what image do you have of a technologist I wonder? The mainstream media regularly refer to science and technology but rarely to engineering. But it is engineers who create technology. For example, the landing of a space probe on a comet in 2014 was widely reported as a scientific achievement. So, with two colleagues, Stuart Burgess and Paul Weaver at the University of Bristol, UK, I sent the following letter to the UK Guardian newspaper. We wrote '*Landing Philae on Comet 67P from the Rosetta probe is a fantastic achievement (One giant heartstopper, 14 November 2014). A tremendous scientific experiment based on wonderful engineering. Engineering is the turning of a dream into a reality. So please give credit where credit is due—to the engineers. The success of the science is yet to be determined, depending on what we find out about the comet. Engineering is not the handmaiden of physics any more than medicine is of biology —all are of equal importance to our futures*'.

For centuries scholars have assumed that engineering is technology and technology is applied science—as I said it is an assumption made routinely by the media. An exception is the philosopher Carl Mitcham[23]. He sees four perspectives—technology as objects, as knowledge, as activity and as an expression of human will. The first three are reasonably obvious. Clearly technical objects are artefacts—the bridges, buildings, computers, space probes and jet engines we mentioned earlier. We should not forget to include artworks and religious works such as paintings, sculpture, and musical instruments. Engineering knowledge is specialised, and it works. Indeed, the success of technology is often quoted by philosophers as evidence of the truth of science. Technological activity includes crafting, inventing, researching, designing, making, operating, maintaining, and decommissioning. Mitcham's final perspective, human will or volition, is less obvious perhaps. But here lies the key to understanding the central differences within STEM (Science, Technology, Engineering and Mathematics). Needing, willing, wanting, desiring, or wishing defines purpose. Confusions about STEM often derive from different uses by different people from different backgrounds in different contexts. We therefore have to be very careful to make clear what we mean.

As I said earlier the purpose of science is to know—by producing 'objects' of theory or 'knowledge.' The purpose of mathematics is clear, unambiguous, and precise reasoning. The purpose of engineering and technology is to produce 'objects' that are useful physical tools with other qualities such as being safe, affordable, and sustainable. All are activities arising from human will that sustains our sense of purpose. Science is an activity of 'knowing' whereas engineering and technology is an activity of 'doing'—but both rely on mathematics as a language and a tool. The methods they adopt to achieve their purposes are so very similar that unless you understand their motive and purpose it is often unclear whether a given person is behaving as a scientist or as an engineer/or a technologist.

But there is a difference between technology and engineering. Like most of the media you may see them as synonymous, or you may see engineering as part of technology. Alternatively, you may see it the other way round—technology as part of engineering. The confusion is rife. MIT (Massachusetts Institute of

Technology) is one of the premier engineering schools in the world. Likewise, until recently Imperial College London was a college of science, technology, and medicine. Now it describes itself as focusing on science, medicine, engineering, and business. Most universities have faculties of engineering and there are many hundreds of professional bodies of engineering such as the UK Royal Academy of Engineering and the USA National Academy of Engineering. Commonly engineers are seen as 'fixers' they come to repair your home heating, your car, or your internet comms systems. Rarely are engineers seen to be the ingenious creators of heart pacemakers, scanners, computers, mobile cell phones, the internet, cars, aeroplanes, space rockets, bridges, buildings as well as the many other things we take for granted such as potable water, electricity, and energy utilities.

Practical life (engineering) and our understanding of our world (science and the arts) leapfrog each other and influence each other in intimate ways. Sometimes one races ahead of the other. In all cases we want certainty, but what we actually get is uncertainty[24]. Uncertainty is embedded in the race because often we want to do things that science is silent about. And in any case as we saw in Chapter 1, science and mathematics are incomplete in very fundamental ways. The key quality that engineers 'bring to the party' lies in the Latin root of the word engineer—ingeniator[25]. Indeed, I have promoted a principle of Ingenuity which says '*value, nurture and develop practical wisdom*'.

The first person to understand the special nature of practical wisdom was Aristotle—he called it *phronesis*[26]—it is an idea that unfortunately, and until recently, got lost in the mists of time. Of course, we cannot simply apply ideas from that long ago, but we can use them to build our own understanding. Aristotle saw five ways of arriving at the truth—he called them art (*ars, techne*), science (*episteme*), intuition (*nous*), wisdom (*sophia*), and practical wisdom—sometimes translated as prudence (*phronesis*). *Ars* or *techne* (from which we get the words art and technical, technique and technology) was concerned with production but not action. The Greeks did not distinguish the fine arts as the work of an inspired individual—that came only after the Renaissance. So *techne* as the modern idea of mere technique or rule-following was only one part of what Aristotle was referring to. *Episteme* (from which we get the word epistemology or knowledge) was, to the Greeks, of necessity and eternal; it was knowledge that cannot come into being or cease to be—it was demonstrable and teachable and depends on first principles. Practical wisdom or *phronesis* was an intellectual virtue of perceiving and understanding in effective ways and acting benevolently and beneficently. It was not an art but necessarily involved ethics, not static but always changing, individual but also social and cultural. Aristotle thought of human activity in three categories *praxis*, *poeisis* (from which we get the word poetry), and *theoria* (contemplation—from which we get the word theory). As I see it, in a modern context, *phronesis* is a means towards an end arrived at through moral virtue. It is concerned with the capacity for determining what is good for both the individual and the community. It is a virtue and a competence, an ability to deliberate rightly about what is good in general, about discerning and judging what is true and right but it excludes specific competences (like deliberating about how to build a bridge or how to make a person healthy). It is purposeful, contextual but not rule-following. It is not routine or even well-trained behavior, but rather intentional conduct based on tacit knowledge and experience, using longer time horizons than usual, and considering more aspects, more ways of knowing, more viewpoints, coupled with an ability to generalize beyond narrow subject areas. *Phronesis* was not considered a science by Aristotle because it is variable and context dependent. It was not an art because it is about action and generically different from production.

Thomas Homer-Dixon[27] wants to close the 'ingenuity gap' between those who adapt well to complex changes and those that don't. He wants to see us move from authoritative top-down command and control to a mature collaborative culture that focuses on enabling others to be successful.

If our new common enemy is climate change then ingenuity and engineering will be crucial in our attempts to rescue the golden goose.

What Do Engineers Assume?

First, they assume the whole of science—but are especially aware of context in which it has been tested and shown to be dependable. Why, because science is the most reliable form of knowledge that we have. But throughout history science has developed and changed in very fundamental ways. The story of engineering naturally divides into five ages—gravity, heat, electromagnetism, information, and complexity. From the ancient skills used to build pyramids to the modern engineering of skyscrapers we have systematically developed our scientific understanding of gravity and used it to build bigger, higher, and longer. Our primitive control of fire has developed into mechanical and chemical power from heat through steam, internal combustion and jet engines and manufactured materials. Electromagnetism is a relative latecomer in the long history of human development which has given us electricity, motors, computers, and telecommunications. Out of this came the age of data and information which has turned now into the age of complex systems.

Systems thinking is helping us to integrate disparate specialisms by seeing tools as physical 'manipulators' of information and energy embedded in 'soft' people systems. From the science of Aristotle to Newton to Einstein, from the craft of Vitruvius to Leonardo to William Morris and from the engineering of Archimedes to Faraday, to Berners-Lee and the world wide web, the story of engineering is racing ahead at an ever-increasing pace. It is a story that has had, and is still having, a profound influence on the quality of human life[28].

ECONOMICS AND ENGINEERING ARE BEDFELLOWS

Towards the end of my academic career, I was invited to join the board of a small utility company. The CEO and the Engineering Director had both trained as civil engineers, so I felt I was amongst friends. The rest of the board was made up of men who all had extensive financial experience, and all were very welcoming. Nearly all of them were retired Finance Directors or previously CEOs of other companies. I was brought in as an obviously independent Non-Executive Director with no affiliations that could be interpreted as conflicts of interest. The Chairman repeatedly stressed that we were a local utility serving a local community for over 150 years and wanted to stay that way.' I felt exposed as I had no business experience except as the holder of a large budget for my university faculty. At my first board meeting I met the man who I was to replace. He said to me '*Remember, David, your first duty is to the shareholders.*' I nodded but felt uncomfortable. I thought to myself—surely my duty is to customers and employees and indeed all stakeholders. But I said nothing. I was unsure of myself and not wanting to be critical at that first meeting. But I will return to that unfortunate emphasis on shareholders at the expense of other stakeholders, including the natural world, in Chapter 7. The shares in the company at that time were not highly traded and held a steady constant value on the FTSE. The yield was solid with many shares own by customers. However, one large shareholder was a pension fund. They made it clear to the CEO and FD (Finance Director) that they wanted to see some profit from trading shares. They put pressure on the board to 'gear up'—something that was popular amongst financiers in the run up to the financial crisis of 2008. They wanted to see the company borrowing more from the banks and the market and less emphasis on financing through shares. At first the board resisted. But the pressure continued and grew stronger, and the board changed its mind and decided to gear up. I was the first to admit I did not understand the argument. The discussions were difficult and financially technical. I could not see any benefit to customers or indeed employees.

In the event, the effects of gearing up were remarkable. The share price rose dramatically, and the company attracted attention from foreign investors. In 2010 the company was bought lock stock and barrel by a Spanish company, supported by the Spanish government. They paid a high premium. There was no impact on customers—indeed many of them didn't notice any change. The local company was no more—the board contained a number of Spaniards, and my tenure was finished. There was a set of unintended consequence I certainly didn't foresee, and I think no other board members foresaw either. Our assumptions and intentions were shattered. Why did this happen? The answer in a nutshell is because that was what the market expected would happen. I saw almost no consideration of the wider set of stakeholders in the deliberations and negotiations.

THINKING ABOUT ECONOMICS

Overall economics is a social science that studies the production, distribution and consumption of goods and services—the material welfare of we humans.

There's an old joke that when two economists fall into a hole and realise, they are trapped then the first step in their plan to get out is...assume we have a ladder. An economic advisor is one handed. Why because he always starts by saying 'on the one hand... and on the other... .'

Economists have to make some assumptions. What are they and how do they affect us? A good person to start with is Adam Smith, who lived from 1723 to

1790 in Scotland. He is generally accepted as the father of classical economics[29]. With his leadership and the contributions of David Ricardo, Thomas Malthus, Jeremy Bentham and J.S. Mill, classical economics flourished in the late 18th and early 19th century. These early thinkers saw the economy as a largely self-regulating system working according to natural laws of production and exchange. The wealth of a nation is not measured, they said, by the amount of gold in the King's coffers as it had been up till then, but rather by its national income. The income was in turn determined by the labour of its people, organized by a division of labour using accumulated capital. They saw a role for the state in benefitting the common good, but they thought that unemployment was simply the result of oversupply.

Figure 1.1 Adam Smith
(Public domain—the Muir portrait en.wikipedia.org)

Adam Smith (Fig. 1.1) postulated three laws of economics which are really assumptions expressed as rules. They are the law of supply and demand, the law of competition, and the law of self-interest. We will look at these laws and what they imply later but first we need to recognize that these are not laws in any legal sense of rules that are recognized and enforced by authority (judicial decision). Whether they are scientific laws is a matter of debate since they are difficult to test. Some economists argue that they are scientific laws in the sense that they relate cause and effect between man and economics. Others, like English economist Alfred Marshall, pointed out that they are not definite, precise, or exact—rather statements of tendencies. Alfred Marshall lived from 1842 to 1924 and was one of the most influential economists of his time and a founder, with others, of what came to be known as neoclassical economics. He introduced utility, developed by Jeremy Bentham and J.S. Mill, into economics. Utilitarianism is a normative ethical theory that says we should act to maximise happiness and wellbeing. He developed theories of supply and demand that were driven by production, consumption, and pricing. He assumed that the production, consumption, and valuation (pricing) of goods and services are driven by Adam Smith's supply and demand but that the value of a good or service is determined by maximizing the utility of a rational person. We shall look at this important notion of utility later in the chapter.

The view that the laws of economics are synthetic a priori reasoning (recall Rosen's Natural Law) was held by Ludwig von Mises (1881–1973) one of the leaders of the Austrian school of economics. He said that one cannot falsify such laws empirically since they are true in themselves—they do not require empirical testing. But this extreme view has been severely criticized as embarrassingly those of an idiosyncratic crank while others have read them as so uncontroversial as to be trivial[30].

Therein lies the economic dilemma—there are so many theories it is difficult to know which ones to rely on. Most economists would agree that economics is about consumption. The exchange of economic goods and services and how they are priced. Consumption is the objective. Production is the means. Inevitably economics has to be concerned with the level of employment both at the level of individual households and at the level of national income and national goods and services. But should the aim of policymakers be to create employment or to create consumption? If the aim is consumption, then it is assumed that prices are ultimately determined by the consumer. If the aim is full employment, how prices are determined is less clear.

There is currently a deep split between mainstream or orthodox economists and those heterodox economists who espouse Modern Monetary Theory (MMT). Some orthodox economists simply ignore MMT—for example, the renowned Director of The UK Institute for Fiscal Studies, Paul Johnson, does not mention it in his new 2023 book 'Follow the Money'[31]. He follows the orthodox interpretation that government spending without commensurate tax rises will fuel inflation. Jerome Powell, USA Federal Reserve Chairman said in 2019 "*The idea that deficits don't matter for countries that can borrow in their own currency I think is just wrong*", when giving congressional testimony in the Senate. He did, however, admit he had not read up on the theory[32]. Martin Wolf, the chief financial correspondent of the Financial Times, in his recent book 'The Crisis of Democratic Capitalism' thinks the theory of MMT carries dangers[33]. It could be used wrongly by non-economist policy makers; it risks losing control of monetary system and create asset bubbles and inflation (which we will discuss in more detail later).

MMTers think that much of orthodox economics has things the wrong way round. Rather than tax first and then spend on government initiatives, they maintain that countries, such as USA, UK, Japan, with their own fiat currency spend first and then collect tax. These countries are not like households because they have the power to issue their own currency. Fiat simple means by decree. Fiat is an important word here. Fiat money is not backed by any physical commodity such as gold but rather by the government who issued it. Fiat currency is a promise from the government to pay. Its value derives from that promise which in turn depends on the stability of that government. The US$, £GB and ¥ Japanese yen are examples. Fiat issuing countries can never go broke and they can always meet their obligations by paying in their own currency. They can also set the interest rate on any obligations they issue whilst allowing the markets to set other rates. MMT does not claim that governments should spend without limit—regard must be paid to the constraints of real resources and so it matters where government directs its spending. If resources are scarce and poorly targeted, then excessive government spending can cause inflation. If the price of imports like oil, increase rapidly then there is inflation.

Critics of MMT claim that taxes and borrowing fund the spending of the Treasury of government. MMTers say that because the government spends first and then collect taxes back then government deficits create a surplus in the private economy. MMTers such as Stephanie Kelton[34] argue that most of the time deficits are too small not too big. National debt is the accumulated deficits. It worries many politicians. Inflation is caused by overspending when supplies aren't sufficient to meet demands. It follows that deficits don't over burden the next generation nor 'crowd out' private investment. Deficits do not make the issuing country dependant on foreigners. Social entitlements are not creating a long-term fiscal crisis. Murphy[34] refers to our national political paranoia about debt that is crippling our society.

MMT exposes some myths around government deficits and debt. It is largely based on accounting practice and the flow of money (see Appendix). As Wynne Godley pointed out orthodox economics has inconsistent stock and flows. He was using the word stock as a special case of potential. MMT uses balance sheets to make sure the stock flows are consistent. The arguments are not straightforward so we will return to them and examine the assumptions made by economists later in the book and in the Appendix.

These are important issues because orthodox economics is holding back the necessary investment in dealing with climate change—as well as other big challenges such as wealth inequality, corruption, and new technology. As Nersisyan and Randall Wray write[35] '... *we can afford the real MEOW. We already have the financial wherewithal needed to afford whatever is technologically possible. We do not need to go hat-in-hand to rich folks to get them to pay for it. We do not have to beggar our grandkids to pay for it. We do not have to borrow from China to pay for it. We do not have to get the Fed to "print money" to pay for it. All we need to do is to remove the self-imposed constraints, the myths, and the misplaced morality; then budget for it, approve the budget, and spend. No new spending process is required. Follow the normal procedures that the Fed and Treasury have developed. That is how you pay for it*'.

BEWARE OF BEING SELFISH—IT CAN LEAD TO GREED

Flies are attracted to a jar of honey. They swarm around it greedily. But then their feet and wings become so smeared with honey that they can't escape. Their greed is their downfall. Capitalism contains a surreptitious assumption of individualism and a tendency towards concern with oneself at the expense of others, i.e., self-centeredness and greed. One example is 'greedflation'[36] where companies exploit inflation to push up prices to make extra profits. Another is that fear and greed are aspects of the herd mentality that causes investors to react inappropriately to events such as the contagion spread by viral social media posts and instant access banking on the collapse of the SVB Bank in the USA to vulnerable banks such as Credit Suisse and others. As Warren Buffett has said[37] "*Investors should remember that excitement and expenses are their enemies. And if they insist on trying to time their participation in equities, they should try to be fearful when others are greedy and greedy only when others are fearful*".

The term economic man, as derived from Adam Smith's law of self-interest, and widely used in economic models, set the thinking of economists in a seriously wrong direction. He wrote[38]. *'It is not from the benevolence of the butcher, the brewer, or the baker, that we expect our dinner, but from their regard to their own self-interest. We address ourselves, not to their humanity but to their self-love, and never talk to them of our own necessities but of their advantages.'* Smith's idea of competition was also central to his thinking. In an open market when those who are willing to pay cannot get what they want then some will be willing to give more. He wrote *'A competition will begin among them, and the market price will rise...'*.

However, Smith wasn't amoral, he didn't have the narrow view that others later attributed to him. He didn't believe in selfishness and greed—that came from his subsequent followers. He believed in fair play and fair distribution of economic rewards[38].

The narrow assumptions about self-interest as selfishness are arguably necessary, but they are by no means sufficient for human flourishing. Orthodox economics, as part of liberal capitalism, assumes that a consumer's first concern is to maximize personal satisfaction. Clearly that is an important motivation that must be recognised. The idea of an economic man is widely used by economist to represent an idealized person who acts rationally, with perfect knowledge and who seeks to maximize personal utility or satisfaction. Rational purchasing decisions do depend on evaluations of the utility of a product or service and market equilibrium does arise from a balance of supply and demand. But the totality of the reality is much more complex. Utility as the usefulness, satisfaction, or enjoyment that a consumer can get from a service or good such as food, clothing, accommodation, car, or TV set, is too narrow a concept for collective human flourishing. Of course, people act in their own selfish interest—but not always as we have recently witnessed strikingly during the Covid crisis and the war in Ukraine with countless acts of altruism. In my view, focusing on selfish interest builds in the pursuit of individual pleasure and encourages self-centered greed. We need to change the values of economic man away from utility towards mutual obligation to each other and to our environment in line with Adam Smith's original intentions. We have obligations to our own interest but also to others and to the natural world. As I said earlier, we need to engage with the citizen story as promoted by Jon Alexander[39] and others.

Selfish interest leads to conspicuous consumption in some. Thorstein Veblen thought that the conspicuous consumption of the very rich was a kind of showing off. For example, women's clothes were purposely impractical to show they never had to peel potatoes or clean a window. He thought that people do not decide rationally but rather to attract approval. But that utility is marginal because it reduces with each increment of additional satisfaction or benefit that a buying consumer gets.

Selfish interest in business leads to a culture of making as much money as possible—to maximize profits. Milton Friedman (1912–2006) is quoted as saying

'*The social responsibility of business is to increase its profits*' He was implying that if lots of people value climate friendly activities, they would choose to spend their money on them. But that begs the question as to whether individuals have enough information and power to make good decisions about what it is ethical to buy.

THINKING ABOUT DEMOCRATIC CAPITALISM

Martin Wolf[40] writes that economic failings have shaken faith in global capitalism. The marriage between capitalism and democracy has become fraught—yet a divorce would be unthinkable. They need each other to survive (Fig. 1.2).

Figure 1.2 Capitalism—Advantages and Disadvantages

He emphasizes the role of citizenship and a shared faith in the common good. Democratic capitalism is justified by politicians and others by saying it gives us freedom of choice, and incentives to be efficient and innovative. It has led to economic growth and increased standards of living in the West. But we should remember that not everyone agrees. In 1974[41] only about 30% of the world's independent states met the criteria of electoral democracy where citizens can choose and replace their leaders in free, fair, and meaningful elections. Larry Diamond[42] comments that up to 2007 the average level of freedom in the world (measured annually by Freedom House[43]) rose slightly and came to an abrupt halt in 2006. Of course, it is not easy to classify regimes as democracies or not. It is a highly uncertain exercise. Freedom House relies on continuous measurement of key variables such as political rights, and civil liberties and using a somewhat arbitrary cutoff point to separate democracies from nondemocracies. Nevertheless, the evidence suggests that since 2006, democracy has declined.

Why is this? Diamond says it is bad governance. The biggest problem in Africa is corruption and abuse of power for private gain. Every region of the world scores worse on transparency and the rule of law than on political rights and civil liberties. Leaders in these countries are eroding democratic checks and balances, removing accountability, overriding term limits, and accumulating power and wealth for themselves and their families, cronies, clients, and parties. In the process, they demonize, intimidate, and victimize, jail or murder opponents who get in their way. Democratic parties and parliaments are weak, and the bureaucracy lacks the expertise and independence to manage the economy. Weak economic performance leads to rising inequality which exacerbates the problems of abuse of power, the rigging of elections, and the violation of the democratic rules of the game.

But there is hope. Diamond reports that most importantly, there has not been significant erosion in public support for democracy. Many Africans, for example, understand the importance of political accountability, transparency, the rule of law, and restraint of power. What is more, they would like to see their governments manifest these virtues. Democracy may be receding somewhat in practice, but it is still globally ascendant in peoples' values and aspirations. This opens up significant new opportunities.

Of course, even the most fervent supporters of democratic capitalism recognize it as not being perfect but, they argue, is better than all other systems. It is failing us in ways that have to be addressed—especially perhaps in the USA. Another Freedom House report[44] identifies three enduring problems that have undermined the health of the US political system: unequal treatment for people of color, the outsized influence of special interests in politics, and partisan polarization. Weaknesses include partisan gerrymandering in elections, polarization through radically partisan positions in primary elections, tying partisan affiliation to demographic traits, and undermining a sense of common national identity. The report concludes that these three problems are creating a vicious circle of distrust and dysfunction. Unless they are adequately addressed Americans' faith not just in their government, but also in democracy itself will continue to decline. Michael J. Abramowitz, president of Freedom House, says[45] "*the strength of American democracy is important for people everywhere, not just here at home. Congress and the Biden administration must make it a priority to strengthen our institutions, restore civic norms, and uphold the promise of universal liberty on which our nation was founded*".

Capitalism tends to produce monopolies with consequent low wages, boom and bust cycles, increasing inequalities and tendencies to be selfish and greedy. In a democracy we elect the politicians based on what they promise us. They appeal directly to our selfishness. We need them to change but we can only do that if we all understand better what is going on. One particularly important example of this is fiat money that we introduced earlier and its role in fiscal and monetary policy which we will look at in more detail in Chapter 4. It is a lack that leads to politicians making bad policies on taxation, managing the supply of money and the passing on of debt to our children and grandchildren.

LEAVING VIOLENCE AND SELFISHNESS BEHIND

In Anglo-Saxon Britain (5th to 11th centuries) the world was full of betrayal, vengeance, and violence. Violence was at the heart of how we came to be governed. Inevitably in that unruly world some people submitted to others—often just to be able to eat and sustain their families. The world lord derives from the old English *hlaford* meaning *'loaf guardian'* or bread giver. New lords coerced their weaker neighbours into subservience and the most successful emerged to become kings. The Anglo-Saxon Tribal Hidage, written in the 7th to 9th centuries, is a list of thirty-five tribes with several independent kingdoms and smaller territories—each of which contained a number of hides. A hide was an area of land that supported an extended family. At that time war between kingdoms was virtually continuous. Eventually seven regions dominated. It wasn't until 927 that they were ruled by one King of England—generally acknowledged to be Aethelstan until 939.

Violence has continued over the centuries to dominate politics and power and ultimately still does as the dreadful events of the 21st century in Afghanistan, Syria and Ukraine demonstrate. People still commit acts of violence on each other. Historically rates of homicides are difficult to establish but data does indicate a decline though variations between countries are large[46.] But over time the incidence of violence and war has declined—largely due to improved literacy and education and the emergence of political freedom through democratic government.

Nevertheless, as the killing of Nauru's golden goose demonstrates, we are still allowing business models built on false economics to ruin the lives of many people around the world. If we treat natural commodities as though they are products then countries like Venezuela, Nigeria, Iraq, Angola, Uganda which are rich in natural resources will continue to suffer from Richard Auty's resource curse referred to earlier. Petroleum, for example, tends to boost authoritarian regimes, increase corruption, and trigger conflict[47]. Resource poor countries like Korea, Taiwan, Hong Kong, Japan, and Singapore have fared much better. We can learn much from them. They have the right basics, efficient legal systems, minimum bureaucratic barriers, innovative investment in education and infrastructure with a capable public administration. Their social norms include good will, hospitality and a sense of community, respecting others, food and where it comes from, nature, and aesthetics.

If we are to move from the consumer story to the citizen story as Jon Alexander promotes—then hope and power will come from how we learn to collaborate and cooperate to release personal agency—to find our way, side by side—democracies and non-democracies alike. We need to believe that we humans are collaborative, creative, caring citizens by nature—and if we give ourselves and each other the chance to save the golden goose. If wars are like squabbling over the deckchairs on the Titanic, then democratic and non-democratic countries have to engage in conversation about our common enemy—climate change.

WE MUST LEARN TO AGREE TO DISAGREE WITHOUT KILLING EACH OTHER

Clearly Russia and China are key players in getting agreement about the single most important issue for humankind—recognizing and acting to combat the effects of climate change. We can agree to disagree about the way we run our societies and governance. We can agree that killing each other is not good for anyone. But we *must agree* on what we need to do to combat climate change. Both countries give us some signs of hope. It is reported that Vladimir Putin is changing his mind[48]. During his two decades in power, Putin has gone from joking about the climate crisis to gradually accepting responsibility for responding to climate change. Russia is the world's fourth-largest greenhouse gas emitter, and it is warming faster than the rest of the planet. Melting permafrost[49] is believed to have caused a tank to split, spilling around 20,000 tons of diesel into the rivers and lakes. Across Russia's frozen north the ground's ability to support buildings will degrade by up to a third by 2050, creating an enormous infrastructure disaster that one study estimates could cost $132 billion. And while Russia pursues its imperial war in the south, in the north, climate change has launched a chemical one. Anthrax released from the melting soil in recent years[50] is only the first warning shot. China is the world's top emitter, producing more than a quarter of the world's annual greenhouse gas emissions, which contribute to climate change. Under the Paris Agreement China has pledged to cut emissions, reduce coal use, and invest in renewable energy. But it still finances coal-fired power plants abroad. Air pollution, water scarcity, and soil contamination remain threats to the health and livelihoods of China's people, increasing dissatisfaction with the government. And all of this is complicated by the continuing Covid problem and its far-reaching consequences. But remember one thing from this chapter—let us not permit our disagreements to prevent common action on climate change.

We killed Nauru's golden goose by being greedy. We must not make the same mistake about life on earth—it is in real danger. To do that we must make many changes. The first is to change the way we think—to recognize that we become what we think. Thinking involves assumptions about all sorts of things. Our personal assumptions help us to live in hope and fulfillment. Professional assumptions set standards of competence. Sometimes the assumptions strengthen our sense of self-worth. They help us to join likeminded groups to enjoy the things we like to do and feel some sense of purpose. Sometimes they protect us from cowboys who pretend to be what they are not. They help us to identify people, such as doctors and dentists, engineers, and economists, who are qualified from those who aren't. Assumptions help us to see and relate to the world we live in. But we have to be careful because assumptions can and often do lead us astray. We attribute the wrong causes to the wrong effects. So, we must become aware of our own assumptions. Social media is changing social norms. Our assumptions help define our social norms, and they depend on our joint expectations. We must become more critical of them. Many of us are losing touch with nature and we need to deliberately go out and experience it. We are part of the natural world.

In science where once we only saw relative simplicity, we now see complexity. Complexity is not an easy idea. Realizing how interdependent things can be mind-bending. The butterfly effect of chaos theory (a special case of complexity), the natural world of mutual interest between plants and animals, including we humans, forces us to understand in new ways. Thinking in terms of systems helps us see the big picture as well as the detail. It helps us go beyond scientific reductionism. Engineering and economics are both essential to modern progress and both are not well understood by non-specialists. Engineering is more than technology because it requires practical wisdom. The word implies ingenuity, and we need to nurture that for human flourishing. Economic thinking is currently deeply split between the orthodox and heterodox of which Modern Monetary Theory is a part. Democratic capitalism needs to change to principled capitalism guided by moral virtues. Profit for shareholders is important but companies now must be aware of their obligations to life on our planet Earth.

The climate crisis has already created many voices for change in the ways we relate to each other. But the old joke about a tourist in Ireland is telling. When he asks a local for directions to Dublin—the reply is: *'Well if I were you, I wouldn't start from here'.* The case for social, economic, and political change can only start from a different place if we examine carefully where that place is. I maintain a good start would be a paradigm shift from the *selfish rational man* to one of the *rational person of obligation.* A change from the consumer story to the citizen story. A change in the social norms of some to eliminate corruption by striving to reinforce the good and reduce the bad in people. It means creating a greater awareness of moral disengagement and understanding why some people act inhumanely without being stressed.

Together we have to save the golden goose. We need to begin to explore how to change capitalism for the better and appreciate the need to embrace essential simplicity within uncomfortable uncertainty and complexity. Across nations with diverse form of governance we must learn to agree to disagree without killing each other.

End Notes

1. The Parable of the Sadhu is adapted from Beyer, J.M. and Nino, D. 1999. 'Ethics and cultures in international business,' Journal of Management Inquiry. 8(3): 287–297. The story describes how each of the hikers were so focused on reaching a holy place, that they ignored their ethical responsibilities to a holy man. The story highlights the tension that can arise between corporate and individual goals.
2. Auty, R. 2002. Sustaining Development in Mineral Economies: The Resource Curse Thesis, Routledge, UK. See also Connell, J. 2006. Nauru: The First Failed Pacific State? The Round Table. 95(383): 47–63. See https://www.tandfonline.com/doi/epdf/10.1080/00358530500379205?needAccess=true&role=button
3. Bicchieri, C. 2016. Norms in the Wild. Oxford University Press, UK. See also https://www.sas.upenn.edu/~cb36/files/chan.pdf

4. Immanuel, Kant (1724–1804) is a central figure in modern philosophy. He brought together rationalism and empiricism. He said that human understanding is the source of the general laws of nature and moral law. Scientific knowledge, morality, and religious belief are mutually consistent and secure because they all rest on the same foundation of human autonomy.
5. Juneman, A., Suleeman, J. and Takwin, B. 2018. Psychological mechanism of corruption. Asian J. Sci. Res. 11: 587–604. See https://scialert.net/abstract/?doi=ajsr.2018.587.604
6. Bandura, A. 2015. Moral Disengagement: How People Do Harm and Live with Themselves. Worth Publishers. For a summary of moral disengagement. See Psychology 2023. Moral Disengagement. See https://psychology.iresearchnet.com/sports-psychology/moral-development/moral-disengagement/
7. Pinkstone, J. 2018. Why 536 AD was the worst year to be alive. MailOnline at https://www.dailymail.co.uk/sciencetech/article-6397621/Why-536-AD-worst-year-alive.html
8. Magee, B. 1973. Popper. Fontana Modern Masters, UK.
9. For examples of silos in Health Care, see Kelly, Y.P., Goodwin, D., Wichmann, L. and Mallika, L. 2019. Breaking Down Health Care Silos. Harvard Business Review at https://hbr.org/2019/07/breaking-down-health-care-silos. Silos in economics see Investopedia at https://www.investopedia.com/terms/s/silo-mentality.asp. Silos in organisations. See https://www.lightercapital.com/blog/what-are-organisational-silos
10. Robert, R. 1990. The modeling relation and natural law. In Mathematics and Science, pp. 183–199. World Scientific Publishing. http://dx.doi.org/10.1142/9789814503488_0013
11. Kurt Gödel's two incompleteness theorems are important in modern logic. They address the limits of provability in formal axiomatic theories. The first theorem states that in any consistent formal system within which a certain amount of arithmetic can be carried out, there are statements of the language of which can neither be proved nor disproved. The second theorem is that a formal system cannot prove itself to be consistent. See https://plato.stanford.edu/entries/goedel-incompleteness/
12. Friends of the Earth, Importance of nature. https://friendsoftheearth.uk/nature/importance-natu species are going extinct and the natural systems that support all life on Earth are being eroded faster than ever before.re
13. Hatty, M.A., Goodwin, D., Smith, L.D.G. and Mavondo, F. (2022, May 12). Speaking of nature: Relationships between how people think about, connect with, and act to protect nature. https://doi.org/10.5751/ES-13369-270317
14. Complexity is cognitively demanding. Cynefin is an example of attempts to come to terms with it in making sense in our understanding of the natural world. It is a decision support framework and not a methodology. See The Cynefin Framework 2023. The Cynefin Company. https://thecynefin.co/about-us/about-cynefin-framework/). It has four domains, (a) clear ordered and simple, (b) complicated, (c) chaotic, and (d) complex. Shifts between domains may be changes of zones or phases such as when a liquid changes to a gas. As human observers we can move dynamically between domains as we try to make sense of what we are experiencing. The differences between domains are defined by the type or absence of constraints. For example, if future outcomes are predictable under sustained constraints, then the domain is clear, ordered, and simple. In the complicated domain, specialist expertise is required. For example, a jet engine is complicated and entirely opaque to a layman. Qualified engineers can design, build, and maintain them successfully. Chaos is the absence of constraint—the border between order and chaos. Chaos is full of instabilities where states of the systems bifurcate and lead to new behaviors. So, a sudden bifurcation or instability causes a large change in the system. This leads to phenomena where small changes in the starting points of a

simulation of a system leads to entirely different futures. The flap of the wings of a butterfly in the UK causing a tornado in Florida is known as the butterfly effect. See Lorenz, E.N. 1963. Deterministic nonperiodic flow. J. Atmospheric Sciences. 20(2): https://journals.ametsoc.org/view/journals/atsc/20/2/1520-0469_1963_020_0130_dnf_2_0_co_2.xml. Chaos is a special case of complexity. The chaotic instabilities mean that it is impossible to trace back past trajectories of the state of a system. Time becomes irreversible. Chains of cause-and-effect are difficult to trace, and small differences may be amplified. In their 'far from equilibrium' state complex systems can be dissipative systems (see Chapter 2 and Note 17) that self-organise as they evolve (see Prigogine, I., Stenger, I. 2017. *Order Out of Chaos* (2nd ed.) Verso). That gives them some resilience rather like that we see in the complexity of the natural world as plants and animals evolve and adapt to changing circumstances or contexts. However, they can also be fragile if a small failure in one holon spreads in a chain reaction of cascading failures. Recent examples are loss of biodiversity, the implosion of the World Trade Centre in New York, known as 9/11 and the banking collapse of 2008.

15. This is taken from a poem by Rudyard Kipling (1902) in The Elephant's Child keep six honest serving-men (They taught me all I knew); Their names are What and Why and When and How and Where and Who.
16. Blockley, D.I. 2020. Building Bridges: Between Theory and Practice. World Scientific, UK.
17. Donne, J. 1623. See https://www.litcharts.com/poetry/john-donne/no-man-is-an-island
18. Koestler, A. 1967. The Ghost in the Machine. Penguin, London. He introduces the term holon to mean a whole (*holos*) and a part (*on*) at the same time. You are a holon since you are a whole individual made up of parts whilst at the same time being a part of the many social groups (not least your family) to which you may belong. Holons are arranged in a holarchy at the top of which is the entire Universe and at the bottom of which is the latest discovery of the fundamental particles of an atom.
19. There are exceptions. One is that the World Economic Forum 2021 asks 'What 'systems thinking' actually means—and why it matters for innovation today'. See https://www.weforum.org/agenda/2021/01/what-systems-thinking-actually-means-and-why-it-matters-today/. See also Cone, E., Reynolds, M. (2020). Beyond the Crisis. See https://resources.oxfordeconomics.com/systems-thinking-program?_gl=1*1heey5w*_ga*MTEzODE1MDAzNC4xNjcwMzI1NTAy
20. Clarke, D. 2014. Poll shows 85% of MPs don't know where money comes from. Positive Money. https://positivemoney.org/2017/10/mp-poll/
21. Magic money tree. See my discussion in Chapter 4.
22. https://insight.ieeeusa.org/articles/engineers-in-congress/
23. Mitcham, C. 2022. Thinking through Technology. University of Chicago Press, USA.
24. Uncertain Times, Unsettled Lives: Shaping our Future in a Transforming World. The Human Development Report 2021/22. UN Development Programme. See https://www.undp.org/publications/undp-annual-report-2021
25. Blockley, D.I. 2012. A Very Short Introduction to Engineering. Oxford University Press, UK.
26. Blockley, D.I. 2014. Practical Wisdom and why we need to value it. See https://blog.oup.com/2014/07/practical-wisdom-vsi/
27. In his book, The Ingenuity Gap. 2000. Thomas Homer-Dixon defines ingenuity as the constituent parts of our social and physical worlds in way that help us achieve our goals. See https://homerdixon.com/writing/books/the-ingenuity-gap/

28. Blockley, D.I. 2012. (*op. cit.*)
29. Adam Smith, A. .1776. An Inquiry into The Wealth of Nations. W Strahan & T Cadell, London.
30. Scheall, S. 2016. What is Extreme about Mises' Extreme Apriorism? Centre for the History of Political Economy. CHOPE, Duke University. Working Paper No. 2016–23, August. See https://hope.econ.duke.edu/sites/hope.econ.duke.edu/files/Mises%27%20Extreme%20Apriorism%20SSRN.pdf
31. Johnson, P. 2023. Follow the Money. Abacus Books, UK.
32. Cox, J. 2019. Powell says economic theory of unlimited borrowing supported by Ocasio-Cortez is just 'wrong'. See https://www.cnbc.com/2019/02/26/fed-chief-says-economic-theory-of-unlimited-borrowing-supported-by-ocasio-cortez-is-just-wrong.html
33. Wolf (*op. cit.*)
34. Kelton, S. 2021. The Deficit Myth. John Murray, is a remarkable book introducing Modern Monetary Theory (MMT). It exposes the myths around government deficits and debt which are holding back investments needed to deal with climate change. First, she shows that the USA federal government is nothing like a household because it has the power to issue US$. Uncle Sam can never go broke. Second, deficits create a surplus in the private economy. She argues that most of the time deficits are too small not too big. Evidence of overspending is inflation when supplies aren't sufficient to meet demands. Third, deficits don't over burden the next generation. Fourth, deficits do not crowd out private investment. Fifth, deficits do not make the USA dependant on foreigners and lastly social entitlements are not propelling the USA to a long-term fiscal crisis. This book is essential reading for anyone interested in how economies work. It is readable, compelling, and convincing. Likewise Richard Murphy's book Money for nothing 2021. The Finance Press available at https://www.taxresearch.org.uk/Blog/wp-content/uploads/2021/04/Money-for-nothing-and-my-Tweets-for-free.pdf is a simple straightforward and clear introduction to macroeconomics and a **must read** for anyone interested in the common misunderstandings of many economists and politicians.
35. Nersisyan, Y. and Randall Wray, L. 2020. How to Pay for the Green New Deal. Levy Economics Institute of Bard College. Working Paper No 931. See https://www.levyinstitute.org/pubs/ppb_148.pdf
36. Inman, P. 2023. Greedflation: are large firms using crises as cover to push up their profits? The Guardian, UK. See https://www.theguardian.com/business/2023/mar/24/greedflation-are-large-firms-using-crises-as-cover-to-push-up-their-profits?ref=biztoc.com. See also Elliott, L. 2023. Bank of England policymaker warns of 'greedflation' risk to consumers. The Guardian, UK. See https://www.theguardian.com/business/2023/mar/07/bank-of-england-policymaker-warns-of-greedflation-risk-to-consumers and Elliott, L. 2023. ECB looking out for price gouging as fears grow over 'greedflation'. The Guardian, UK. See https://www.theguardian.com/business/2023/mar/02/ecb-looking-out-for-price-gouging-greedflation-price-rises-eurozone-inflation-profit-margins#:~:text=ECB%20looking%20out%20for%20price%20gouging%-20as%20fears%20grow%20over%20%27greedflation%27,-Concerns%20that%20a&text=Fears%20that%20Europe%27s%20companies%20are,potential%20price%20gouging%20of%20consumers.
37. Berkshire Hathaway Inc. 2004. Annual Report. See https://www.berkshirehathaway.com/2004ar/2004ar.pdf
38. Smith, A. (*op. cit.*). In his Theory of Moral Sentiments. 1759. Penguin Books 2009. UK. He says that the surest route to happiness is to have gained the respect and good opinion

of others. It is a mistake to equate his interpretation of self-interest with maximising profit. See also Wight, J.B. 2005. Adam Smith and Greed. Journal of Private Enterprise. 21(1): 46–58. Who says the maxim that greed is good does not represent Adam Smith's views—nor is it defensible in economics. See https://scholarship.richmond.edu/cgi/viewcontent.cgi?article=1030&context=economics-faculty-publications

39. Alexander, J. 2022. Citizens: Why the Key to Fixing Everything is All of Us. Canbury, UK.
40. Wolf (*op, cit.*)
41. Repucci, S. and Slipowitz, A., Freedom in the World 2022 The Global Expansion of Authoritarian Rule. Freedom House. See https://freedomhouse.org/report/freedom-world/2022/global-expansion-authoritarian-rule. See also Freedom House 2023. Countries and Territories: Global freedom scores at https://freedomhouse.org/countries/freedom-world/scores
42. Diamond, L. 2015. Facing up to the democratic recession. Journal Democracy. 26: 1
43. Freedom House Freedom on the Net Research Methodology. https://freedomhouse.org/reports/freedom-net/freedom-net-research-methodology.
44. Freedom House 2021. Special Report From Crisis to Reform: A Call to Strengthen America's Battered Democracy. See https://freedomhouse.org/report/special-report/2021/crisis-reform-call-strengthen-americas-battered-democracy
45. Freedom House 2021. US Democracy Has Declined Significantly in the Past Decade. Reforms Urgently Needed. https://freedomhouse.org/article/new-report-us-democracy-has-declined-significantly-past-decade-reforms-urgently-needed
46. Roser, M. and Ritchie, H. 2013. Homicides. See https://ourworldindata.org/homicides
47. Ross, M.L. 2015. What Have We Learned about the Resource Curse? Annual Review of Political Science. 18: 239–259. See https://doi.org/10.1146/annurev-polisci-052213-040359
48. Moscow Times 2021. Scepticism to Acceptance: How Putin's Views on Climate Change Evolved Over the Years. See https://www.themoscowtimes.com/2021/07/01/skepticism-to-acceptance-how-putins-views-on-climate-change-evolved-over-the-years-a74391. Jakobsen, S.E. (2022). Have Russian climate scientists convinced Vladimir Putin that climate change is real? *Sciencenorway.no* https://sciencenorway.no/climate-policy-russia-society-and-culture/have-russian-climate-scientists-convinced-vladimir-putin-that-climate-change-is-real/1956249
49. Antonova, M. 2020. Science Alert Environment. Russia Is in a State of Emergency After 20,000 Tonnes of Diesel Fuel Leak into River. Science Alert. Environment. See https://www.sciencealert.com/russia-declares-state-of-emergency-after-20-000-litres-of-diesel-fuel-leak-into-river
50. Chognot, J.P. 2020. As Permafrost Melts It's Unleashing Ancient Viruses, Carbon—And Now Fuel Spills. Science Alert. Environment. https://www.sciencealert.com/as-permafrost-melts-ancient-viruses-and-now-fuel-spills-are-being-unleashed

Chapter 2

Models are Cool—There is no Getting Around It

The Wise Doubt—Judge to Make the Complex Simple and not the Simple Complex

AN AESOP FABLE

Lion lay asleep in the forest, his great head resting on his paws. A timid mouse came upon him unexpectedly, and in her fright ran across Lion's nose. Lion laid his huge paw angrily on the tiny creature to kill her. "Spare me!" begged the poor mouse. "Please let me go and someday I will surely repay you." Lion was much amused to think that a mouse could ever help him—but feeling generous he let mouse go. Some days later, while stalking his prey in the forest, Lion got caught in a hunter's net. He filled the forest with his angry roaring as he tried to get free. Mouse recognized the voice and quickly found him. She gnawed through one of the great ropes that bound him and soon Lion was free. "You laughed when I said I would repay you", said Mouse. "Now you see that even a mouse can help a lion."

Man: *"When I was young, I made some simple mistakes."*

Woman: *"Now you are older you make complex ones!"*

MODELS ARE COOL

Once upon a time I thought models were simply toy trains and boats. I remember being given a small model yacht one Christmas. I couldn't wait to try it out on a local pond. The pond was quite large and fortunately for me quite shallow. My dad and I launched the yacht in full sail and off it went, to my delight, to the middle of the pond. But there it stayed—becalmed and not moving. I remember the horror—but my dad was up to it—he took off his shoes and sock and waded out to retrieve it.

Like many kids, I built cars and mechanical devices like cranes with Meccano. It was a set of reusable metal strips, plates, angle girders, wheels, axles, and gears that were connected by nuts and bolts. Nowadays most kids build with Lego bricks.

Model building is a superb hobby for many and a career for some. Grownups build models—some quite elaborate models of buildings and ships and all sorts, some using only matchsticks. Modern model aircraft and cars can be remote controlled and are great fun. The levels of skill required can vary from child's play to professionals making 3D models to order.

Model building is also a profession. Models are built for film making and to commemorate historic events. Professionally construction architects have small scale models of new building developments built to show off their ideas to clients.

But, as I hinted in Chapter 1, models are also much more than 3D physical representations. They are also basic to science and indeed all forms of knowledge. But these are not physical models—these are theoretical models. The idea though, is not without controversy[1] since it involves deep philosophical issues about the nature of reality.

But at its simplest, a model is a representation of something.

For example, what have Sherlock Holmes and atoms got in common? The answer is that they are both models—one of literary fiction and the other of science. But is a scientific model a law, theory, hypothesis, a conjecture, or a system? The differences are subtle and often the words are used interchangeably.

Almost everything you rely on to live your life depends on models. Models are used to create electricity, to supply water, to build bridges, to power computers, to supply food and to run businesses. Models are cool—acceptable, admirable, useful, and fundamental. But there is a drawback. Models are incomplete in ways you may not realise. They express assumptions which may be hidden from view unless you look very closely. Rather obviously a toy car does not normally have an engine. Less obviously the scientific theory we use to build big bridges relies on Newton's Laws. More obscurely quantum mechanics assumes field theory. It is all too easy to either not know that assumptions are being made or to neglect them in some way. If we do that and consequently apply a model out of context there will often be harmful consequences. For example, in 1954 two de Havilland Comet aeroplanes disintegrated in mid-flight. The planes had square windows. It was assumed that this wasn't an issue. However, the large concentrations of stress at the corners eventually caused the metal to crack due to metal fatigue. Consequently, the pressurized cabins disintegrated in flight killing 56 people. Now all aeroplane windows are curved. We learned the lesson.

In Chapter 1 we said that a system is something, discriminated from its surroundings, as a set of models of our reality created by people for reasons which give meaning to and determine the purpose of the system. As I have said I believe that even the deepest laws of physics are models, and each model relies on a set of assumptions—in other words, they are contextual. Scientific and engineering models of physical systems are highly testable in those contexts. They have brought great progress but also lots of unintended consequences such as pollution. Social scientific models, economic models and models of engineering social systems are not so highly tested. Often, we have to rely on our interpretation of the history of their use and test them through critical discussion. Inevitably they are not as dependable as those of classical physics. We may mathematise them, particularly in economics, but as impressive as the mathematics seems to be, the conclusions are only as good as the models themselves. Of course, the models evolve through the participation, sharing and collaboration of many different people over time. In that sense they are cool (acceptable and admirable) but when we use them knowledge for practical purposes, we must be aware of context and be mindful of when that context is no longer appropriate. One of the quickest ways to get into trouble is to use knowledge out of context.

So where did these scientific and economic models come from, what is their provenance, how did they develop and how can we test them. Only when we know that can we examine the pedigree, status, and practical usefulness of the models of today and modify them to manage the effects of climate change.

You will recall Adam Smith's laws of economics and Robert Rosen's Natural Law in Chapter 1. We said that these laws are not legal rules enforced by judicial decisions but rather that they attempt to relate cause and effect. A scientific law is a set of principles, rules and assumptions established by the authority of general agreement between scientists who have set out to test the law as thoroughly as they can. A theory is an explanation of a phenomenon and may or may not be well tested. A hypothesis is a provisional conjecture, opinion, or speculation. Professionals tend to use the word model to cover all these situations—although some may still argue about the subtle distinctions such as a model is an interpretation of a general law. For example, until the recent developments in applications of quantum theory[2], almost all earthbound engineering is based on Newtons three laws[3]. As noted earlier structural engineers apply Newton's Laws to find the forces in a long span suspension bridge such as Golden Gate in San Francisco[4]. They build a theoretical model of that particular bridge and decide if the bridge is safe or requires attention.

Theoretical models therefore come in all sorts of shapes and sizes. As representations they may be used to interpret, explain, probe, predict all kinds of phenomena including the natural physical world (living and non-living), and the human economic and social world. Models are central to economics and engineering, the environment, and the challenges of climate change.

The models that are relied on by engineers are typically considered as two quite separate systems—the physical and the social. The education of professional engineers focusses almost entirely on the former. But when those same engineers go into practice their problems stem strongly from the latter. The physical systems are those things and systems of things that engineers build such as bridges, buildings, computers, jet engines, airports, transport and utility networks, the internet and many more of the goods we rely on every day. The social systems are the people and organisations required to plan, design, manufacture, operate, maintain, and eventually dispose of the physical systems. Engineering is done to fulfill a purpose—human flourishing—but it is done within an environmental, economic, political, and social context.

A VERY BRIEF HISTORY OF CAPITALISM

If we are to justify capitalism, address its problems and evolve it into a form of principled capitalism (based on moral virtues) then we need to understand a bit of its history. To appreciate why our ideas are as they are, and our understanding is as it is, and why we do the things in the ways that we do, we need to understand a little bit of how they came to be. Capitalism is an economic system in which private individuals or businesses own capital goods. The production of goods and services is based on supply and demand in the general market—known as a market economy—rather than through central planning—known as a planned economy or command economy (as in Russia and China). Put more simply capitalism is an economic system where financial entities lend money to other financial entities for a purpose. Why can they do that? As we shall see it is because ever since the first agriculture, we have had surpluses.

For a brief overview I will divide the history of economic thought and practice in 3 stages—ancient (up to the Renaissance), premodern (Renaissance to 18th century) and modern (up to the present). Within those stages are many variants and it is not necessary or possible to cover them all. I shall just briefly cover some of the main points.

Ancient

When people were hunter gathers, they pretty much shared whatever was caught or gathered between immediate and extended family. Later they divided it all within the clan, tribe, or village. As they began to farm in more settled communities people specialized, depending on their abilities, talent, and context. They started to barter. Agriculture and farming created surpluses that eventually were exchanged in markets in villages, and later in towns and cities. As Greek politician and economist Yanis Varoufakis[1] describes surpluses gave rise to writing because there was a need to keep records of transactions. Surpluses also led to debt as the opposite of surplus because of inevitable delays in the exchanging of goods. Likewise surplus led to money as a proxy for the goods to be exchanged. Finally, surplus led to the existence of the state because of the need for an authority to make the money trustworthy (for more on this see Chapter 5). As one might expect historians still debate how the economy of the ancient world worked.

In parallel to the development of a written language, markings on clay tablets, papyrus and other materials were used in the Bronze Age to account for crops, livestock, and land. Clay tablets have been found in Sumer, Babylonia, the Indus, Chang Jiang, and Nile River valleys. From around 8,000 BCE crops were grown, harvested, and stored in the Nile Valley in Egypt but the state took some as taxes. Wheat, barley, and cooking oil were used to buy other things. Coins began to be used around 500 BCE. The unit of weight of gold was called a deben and was used in bartering. There were roughly a dozen Sumerian city states that traded using the shekel which was a certain weight of barley. The Babylonians used a mixture of commodities fixed by a legal code. The codes included rules for inheritances, taxes, and fines for wrongdoing. The Ptolemaic dynasty of Egypt had a large bureaucracy that kept detailed records on papyri about taxes, government-controlled lands, and labor. The Greek poet Hesiod is credited with the first ever work of economics in a poem written to lecture his apparently foolish brother—saying that hard work is a moral virtue, the world has finite resources—scarcity—and people must make choices. The allocation of scarce resources was also a moral issue for Aristotle. He wrote in his Politics Book1[5] '*Every state is... a sort of partnership, and every partnership is formed with a view to some good... . It is therefore evident that... the partnership that is the most supreme of all and includes all the others ...is the partnership entitled the state, the political association*'. [1252b 1] *The partnership... of several villages is the city-state... exists for the good life*'. Aristotle thought that commerce—the buying and selling of things—to meet one's need was enough. He disapproved of making a profit—gaining from someone else's loss. The ancient Greeks produced and traded goods both locally and long distance and used money to do it, but it seems we know little about the who, why, and how of the trading, what the costs were and how many middlemen were involved. Practice changed and developed over the centuries so what follows is a brief overview. Recent research[6] has revealed that ancient Greece had a market economy some 3,000 years earlier than previously thought with a major trade expansion several centuries before the conquest by Rome. Cereal, olive, and vine pollen were used as proxies for structural changes in agricultural production. Long distance trading was mainly in luxury items, such as gold, silver, jewelry, and fine pottery. Scholars think that the economy was at first embedded in social and political institutions and only later developed a market system—but one unlike modern capitalism. There was no network of interconnected markets so there was no price-setting. The ancient Greek word *oikonomia* is the root of our modern English word economy, but with a very different meaning—*oikonomia* referred to household management. The Greeks valued the wellbeing of the community over that of the individual. Economic activity was about providing sustenance for the family by farming a small plot of land. Land was largely privately owned though much was owned by royal dynasties with peasants who were sometimes slaves working on other royal projects. So male citizens were expected to contribute to the wellbeing of the community by being part of public religious, political, and military life. Economic activities that were not part of feeding the family were of low esteem and labelled *banausic* work meaning mundane and technical. This included manufacturing, business, and trade (that which was not part of the family farm), and any kind of capitalism in the form of investing money to make more money. However, the Greeks did develop banking and credit institutions for changing money and securing deposits in cash and other assets. Banausic work was incompatible with public life and even held as unnatural and immoral. So, profit and capital growth were not approved of. Cities were simply places for people to live and to be centres of religion and government. Cities took the surplus produce of the countryside, manufactured a limited amount of goods, and provided marketplaces and ports of trade. The elite of society were landowners and they dominated politics. Individual city-states certainly had marketplaces largely isolated with minimal connections between them. Governments of the Greek city-states were mainly concerned with maintaining harmony by making laws, adjudicating disputes, and protecting private property rights, making sure that food was reasonably priced and taxing to pay for government expenses. Athens enforced private property rights with officials and law courts. There were officials for weights, measures, and coins to prevent cheating. Athens had laws to prevent the export of grain and to grain imports. Taxes were indirect such as on markets, ports, import-exports on foreigners.

Homer's epic works refer to specialized craftsmen working metal and painted pottery. Mining was important to the economy. Precious metals were used in jewelry, art, and coinage. The value of the coin was directly related to the value of the precious metal (usually silver) with a small mark-up. Mining in Athens was specialized and extensive and could be called an industry—almost in the modern sense of the word. Buildings varied from private houses to monumental stone temples. Houses were unbaked mud brick on a stone foundation with thatched or tiled roof. Temples required organisation, resources, and skill. Work was often contracted to people who either worked alone or who employed others to quarry and transport materials and to sculpt. However, working for wages was looked down on as it was a restriction of freedom, akin to slavery.

Ancient Rome was an agrarian and slave-based economy. The main aim was to feed everyone. Any industrial production was small scale. The need to get grain was one of the factors behind the Roman conquests. Extensive trade routes were established on land and sea. Roman roads are still used today. Goods were traded by barter, but the Romans had a well-developed coinage system. Coins of brass, bronze, copper, silver, and gold were minted with strict rules for weights, sizes, value, and metal composition. Roman numerals were used to count.

Middle Eastern countries, India and China had written guides of best economic practices and norms in the first millennium BCE. In the 4th century BCE Indian polymath Chanakya (or Kauṭilya) anticipated classical economic thought by some 2,000 years in his ideas about international trade, taxation, and a labour theory of value. He described how in an autocracy an efficient state can be managed. He discusses the ethics of economics and the duties of a king. He described a system of coinage that had made barter of secondary importance. He listed sources of state revenues as taxes and fines expended on an army, a bureaucracy, social security measures and the upkeep of the royal court.

The 4th century BCE Guanzi essays[7] described supply and demand pricing and the importance of managing the supply of money and a stable currency. Early Muslim scholars such as Ibn Khaldun[8] in the 14th century appreciated the importance of supply and demand in determining prices, the division of labor, social cohesion and thought that population growth depended on wealth. He was one of the first to examine the division of labor, the profit motive, and international trade. He analyzed the perils of monopolies, the benefits of division of labor and the profit motive, and the rise and fall of economic empires. He wrote that the indicators of fall seem to be greed for luxury, corruption of morality, and tyranny. He advocated the integration of economic growth with social, and cultural values, people centric, based on human rights with strong legitimate participative government, good governance and the rule of law, justice, and morality. He thought the state should be neither laissez faire nor totalitarian. Rather it should serve to accelerate human development and wellbeing.

St Augustine of Hippo was born in 354 as the Roman Empire was dying. He preached that the best life was to be had by giving up all possessions and many followed his teaching and became monks. People should not love possessions, he said, rather they were needed only to live a good holy life. After the fall of the Roman Empire trading collapsed, Roman infrastructure decayed, and people grew food for themselves. In England, the various warring kings of regions such as Wessex, Mercia and Northumbria fought for dominance. They had to defend themselves from constant Scandinavian invasions with fighters and funds from their nobles. The kings rewarded their loyal nobles with gifts of land and individual land ownership vanished. The Roman manorial systems, where workers were protected while living on large states, was gradually replaced by 'feudalism'. The system was based not on money but on religion and the relations between rulers and ruled. There was a strict pecking order with God at the top and peasants at the bottom. Augustine of Canterbury (born around 604) was a Roman monk sent by the Pope Gregory to convert the Kingdom of Kent to Christianity. He succeeded and became the first Archbishop of Canterbury—effectively founding the English Church with missionary preaching which eventually spread across Britain. Later Thomas Aquinas wrote that it was fine to make a profit as long as it was put to good use—for example by giving some of it to the poor. He asked himself what was a 'just price' for something. He thought it wasn't the highest price but one that was honest and without any form of cheating. Following Thomas Aquinas, the medieval church condemned usuary as a sin—the lending of money with interest—moneylender would go to hell with thieves and robbers. The best life is to be found by giving up possessions.

In England after the Norman conquest the regions were largely united under one king, William I, and the feudal relationship between king and noble and between noble and vassal took root. Feudalism governance and economics continued throughout the medieval period—the wealthy prospered, and the poor worked the land. Engineering and technology were a crucial contribution[9]. Horse drawn ploughs with newly developed harnesses were introduced to help farmers. Waterwheels began to power mills to grind corn and towns came back to life. Communities started to trade more and as it flourished so did finance. Eventually feudalism diminished as government became more centralised. It was not finally abolished until the 17th century. Feudalism was a social and economic system based on inherited social rank that granted privileges and obligations. It was static, stratified, and exclusive with hundreds of virtually independent feudal lords. Static in the sense that change wasn't easy, stratified from the king at the top and country knights and peasants at the bottom and exclusive since contact with others was hardly needed. Wealth came from agriculture, not through markets but through labour of the serfs to the lords. Serfs paid rents, provided food, worked the land owned by the lord, and helped defend the land from intruders, violence, and warfare. Feudalism was a way of structuring society around reciprocal legal and military obligations between lords, vassals, and the land (fief). The more land owned by the lord the wealthier he was.

Early modern

Feudalism lasted for five centuries but it was local and rigid. It slowly gave way to the more dynamic mercantilism through the rise of nations states and trade with Europe and indeed the world. The gradual change from feudalism to mercantilism (the first form of capitalism) began at the time of the Crusades and was more or less complete by the end of the Renaissance period. The barter system became outdated with coins and bills of exchange taking over. Some peasants left the land to work in cities for money. Kings as absolute monarchs were becoming stronger to the cost of the feudal lords. By encouraging trade, the king increased his personal wealth as well as that of the state. New discoveries of Indian and Atlantic routes, exploring the New World led to a large expansion of commerce. The lure of bigger revenues inspired emperors and kings to patronize expeditions to find new trade routes. Shiploads of gold and silver brought back from South America by the conquistadores to Spain showed the power of the discoveries. Then came the infamous slave trade from the 16th to 19th centuries—abolished by the British Parliament in 1807 followed by the USA—but not internally until the 1860s.

Banking

Banking began around 2000 BCE when grain loans were made to farmers and traders, but it was in the Renaissance period when the form of banking that we might recognize today was established in Italy[10]. Perhaps the most famous was the Medici Bank established in 1397. The first Italian banks were merchant banks based on the Lombard plains cereal crops. Many Jews, who were not allowed to own land, set up benches to trade in crops. Christians and Muslims were strictly forbidden the sin of usury—lending and receiving interest—giving the Jews and advantage. To get round the ban on usuary some Christians offered loans without interest but required insurance instead. They secured rights on the sales of grain and profited from discounted rates against future prices. They financed crop growing and underwrote insurance to guarantee delivery. They also arranged, if crops failed, supplies through alternative sources, and helped to keep farmers in business during a drought. Merchant banking developed from the financing of trade to the holding of deposits of *billettes* or notes. The merchant's benches (banca as in a counter is the Italian for bench from which we get bank) in the grain markets held money against a *billette*, a bill, note, a letter of formal exchange, and later a bill of exchange and then a cheque. The term bankrupt derives from the Italian *banca rotta*, or broken bench—when traders' deposits were lost. Hence the expression of being broke.

At medieval trade fairs moneychangers began to issue documents redeemable at other fairs for currency. If redeemable at a future date, they would often be discounted by an amount comparable to a rate of interest. These eventually became bills of exchange, which could be redeemed at any office of the issuing banker. Now it was possible to transfer large sums of money without moving large chests of gold that required armed guards. Around 1400 these new methods of banking became politically unpopular, and bankers were expelled from some countries. Nevertheless, a number of banking family groups arose in the Netherlands, Germany, Spain and even the Ottoman Empire. In England there were no banking houses until the 17th century even though the London Royal Exchange was created in 1565.

Mercantilism

Mercantilism encouraged regulation by government to increase state power at the expense of rival states. It was a kind of economic absolutism. Ideas were disparate but any gain by one country would be a loss for another. So, every piece of ground was used for agriculture, mining, or manufacture with a large working population in domestic manufacture. Wages were limited and domestic consumption restricted. Exporting gold and silver was banned. In short mercantilism was protectionist so that governments and merchants became partners in increasing political power and private wealth to the exclusion of other states—a form of imperialism. Trade surpluses were for the benefit of government. Mercantilism helped transform Britain into the world's dominant trading power.

Adam Smith was its foremost critic. He said that it confused wealth with money, and that it is consumption that is important. Mercantilism gradually gave way to classical economic thought so that by the 19th century the British government embraced free trade. This happened on different time scales—for example, in France triggered by the French Revolution.

An 18th century group of French economists were called the physiocrats. They emphasized productive work as the source of national wealth in contrast to mercantilism—which had tended to focus on the wealth of the ruler as an accumulation of gold, or balance of trade. The mercantilists believed that value of the products of society was created at the point of sale. It was then that the seller exchanged his products for more money than the products had previously been worth. The physiocrats saw labour as the sole source of value but for them that labour was only agricultural. They thought that all industrial and non-agricultural work was an unproductive appendage to agricultural labor.

Modern

Earlier I referred to the 18th century, Scottish economist Adam Smith as the father of economics. He worked with the ideas of French Enlightenment to develop his views on how economies should work. In the 19th century, Thomas Malthus and Karl Marx expanded on Smith's ideas and then in the late-19th century economists such as Léon Walras and Alfred Marshall began to use statistics and mathematics. Later still John Maynard Keynes developed theories in the early 20th century that are still very influential in the management of monetary policy today. Indeed, most modern economic theories are based on, or in reaction to, the work of Keynes—for example, the free-market theories of Milton Friedman. More recent theories include those of Harvard University economist Amartya Sen who argues for factoring ethics into social welfare.

Adam Smith did not like mercantilism. He began his studies by looking at the work of the physiocrats, Quesnay, and Mirabeau[11]. He took many of their ideas and expanded them into a theory of how economies ought to work, as opposed to how they do work. He concluded that competition was self-regulating. Governments should take no part in business. There should be no tariffs, taxes, or other economic interventions unless they were to protect free-market competition. In his 1776 book "The Wealth of Nations" he set out some of the mechanisms of capitalist production, free markets, and value. He argued that individuals acting in their own self-interest could, as if guided by an "invisible hand," create social and economic stability and prosperity for all. The idea of the invisible hand is still debated today but is often narrowly determined as maximizing profit rather than as Smith proposed acting within a moral framework that derives from our common humanity. To Smith, self-interest means looking after your own advantage and security. Greed only comes about when you consider only your own advantage and disregard the needs and rights of others.

However even the most loyal supporters of Smith's ideas recognize that some of his theories have not aged well. For example, he promoted productive labour, such as manufacturing, as superior to unproductive labour, such as servitude. Clearly wrong in today's service-dominated economy. Smith's labour theory of value—that the value of a good can be measured by the hours of labour needed to produce it has largely been abandoned.

Karl Marx and Thomas Malthus did not like Smith's treatise. Malthus (1766–1834) was grappling with the challenges of the emerging capitalism after the French Revolution and the demands of a rising middle class. Jean-Baptiste Say (1767–1832), David Ricardo (1772–1823), and John Stuart Mill (1806-1873) were also influential around that time.

Malthus predicted that population growth would outstrip food supply. He was wrong because innovative engineering enabled increased production, but he did move the focus of economics onto the scarcity of goods.

Marx declared that the means of production were the most important factors in an economy. He became convinced that capitalism was an unstable part of a class war. However, he underestimated the flexibility of capitalism. Instead of a clear division between owners and workers, the market economy became a mixed class that included both owners and workers. Marx did predict however that businesses would grow larger and more powerful under free-market capitalism if they were allowed to—and that is what has happened.

The Marginal Revolution

English economist William Stanley Jevons (1835–1882), Austrian economist Carl Menger (1840–1921), and French economist Léon Walras (1834–1910) independently developed a new perspective in economics known as marginalism. They had become interested in how markets operate, and market prices fixed.

Their key idea was that people make their decisions around the specifics of what they choose to buy, produce, or sell. They balance the scarcity of each good against the value of its use at the margin. So, the price of a diamond is higher than the price of a flagon of water. They reasoned that although water is a basic need, there is plenty of it and though diamonds are purely decorative they are scarce.

Walras used mathematics to develop a theory of marginal analysis. Out of his work arose a theory of general equilibrium. Economics moved from being purely discursive to one backed up by statistics and mathematics. Then Alfred Marshall (1842–1924) developed mathematical modeling of economies and introduced many concepts that are still not widely understood, such as economies of scale (where average costs decrease with the scale of production), marginal utility (the additional utility from buying an additional unit of something) and the neoclassical paradigm (people make a rational calculation of their own best interests).

In Chapter 1, we saw the importance of the testability of a model in distinguishing science from non-science. It seems to be virtually impossible to do controlled experiments to test economic models. The underlying reason is that economic behavior contains multiple cases of human intentionality—the quality of being done for a purpose or intent. That makes it, like all social sciences, difficult to test and hence predict. Mathematical modeling helps. That is why, in order to simplify some of the complexities of human intention, Walras, Marshall, and their successors assumed rational actors and efficient markets. It is also why statistical methods can be applied to economic data (known as econometrics).

A tenet of mainstream economic thought was that the economy would automatically revert to a state of general equilibrium. The needs of consumers are always greater than the capacity of the producers to satisfy those needs, so that everything produced would eventually be consumed once the appropriate price was found. Then came the work of John Maynard Keynes (1883–1946) which became so important that a whole school of thought, known as Keynesian economics, was named after him. Many see his ideas as the foundation of macroeconomics or the economics of the whole at a large scale big-picture level. During and after the Great Depression of the 1930s he challenged the idea that free markets would create full employment if wages could be managed flexibly. He thought that aggregate demand (total spending) decided the level of economic activity and consequently that poor demand would lead to high unemployment. Keynes used large scale aggregates instead of marginal units or specific goods markets and prices to model the rate of unemployment, aggregate demand, or average price-level inflation for all goods. Keynesian economists regard aggregate demand as volatile and unstable. Consequently, when demand is low, we get a recession and when demand is high, we get inflation. These fluctuations can be mitigated by the policies coordinated between government and central bank. Fiscal policies are government actions and monetary policies are operated by the central bank. If done well they help to stabilize economic output, inflation, and unemployment over the business cycle. In other words, Keynesian economists advocate a regulated market economy, predominantly private sector, but with an active role for government intervention during recessions and depressions.

In the 1950s–1970s microeconomics and macroeconomics dominated economic thought in the Western world. It was known as the neoclassical synthesis and widely taught in universities and practiced by researchers and policymakers. Within this synthesis various channels have evolved—sometimes in direct opposition. The basic tension is between seeing free markets as efficient and beneficial or as inherently prone to calamitous failure. Persistent academic and public policy disagreements followed.

Perhaps the most prominent recent model is that of monetarism and the Chicago School, developed by Milton Friedman (1912–2006). This shifted the emphasis from fiscal policy (favored by Keynes) to monetary policy. Monetarist models assert that variations in the money supply are major influences on national output and price levels. Policies should target the growth rate of the money supply. Monetarism was widely espoused through the 1980s, 1990s, and 2000s. Other models are behavioral economics and Austrian-school economics.

Behavioral Economics was a reaction to the idea that consumers are rational actors who behave in their own best interests. Obviously, people often do not act in their own best material interests but are shaped by non-material psychological factors such as prejudice, rational ignorance and partial information, knowledge, and bias. Naturally, this makes modeling and forecasting more difficult than ever. Some of the new concepts included, the sunk cost fallacy where people continue to invest in a failing project because they are already deeply involved. Another was the availability heuristics or thinking an outcome is more likely because it is more obvious. Bounded rationality is where people act with insufficient information when they know that they could get more.

Amartya Sen[12] has proposed a moral capability framework in which social arrangements should be evaluated primarily according to the extent of freedom people have to promote as well as achieving functions they value. The freedom to achieve well-being is primary but different people have different capabilities to do and be what they value. Sen sees the over-accumulation of wealth by individuals or groups as ultimately harmful to society. Martha[12] Nussbaum has listed some human capabilities for human flourishing as life, health, bodily integrity, senses, imagination and thought, emotions, practical reason, affiliation with others, play and control over one's environment. She sees her list as neutral with respect to the 'good', but one that has some consensus.

SOME ECONOMIC MODELS

Economists try to identify patterns in the growth or otherwise of an economy over time, shifts in types of activity for example from agriculture to industry, population movements from country to city dwelling, changes in engineering technology, for example, the rise of the steam engine or the use of IT (Information Technology) and robotics, and international trade.

Here are just a few of the many laws, derived from Adam Smith's early work, which are quoted by different economists as being fundamental.

(a) The law of demand—the quantity of a good or service bought varies inversely with price. In other words, the higher the price, the lower the quantity demanded. Likewise, the law of supply says that as the price of an item goes up, suppliers will attempt to maximize their profits by increasing the number of items for sale. As we already noted, Adam Smith referred to these laws as an 'invisible hand' that guides a free market.

(b) The law of diminishing returns—the idea that after some threshold, more production will yield smaller returns.

(c) Keynes' psychological law of consumption—increased consumption is less than increased income.

(d) The law of comparative advantage—a company or whole economy can produce a specific good or service at a lower opportunity cost than others. Opportunity cost is the lost benefit that would have been created from an option not taken. In other words that which you must give up obtaining something. For example, if you go on holiday instead of buying a new car the opportunity cost is not getting the new car. This law is assumed to be the basis of mutually benefit through voluntary cooperative trading.

(e) The law of returns to scale—when there is a proportionate change in the amount of input, the behavior of output also changes. For example, if the input is doubled and as a result the output is more than doubled, then that is called an increasing 'returns to scale'. All that means is that the cost per unit is lower which economists call a higher economy of scale.

(f) The law of variable proportion—an increase in some inputs relative to other fixed inputs will after a point, result in the extra output reducing.

SCARCITY IS ONE IMPORTANT FACTOR

As a small boy, just after World War 2, my mother used to occasionally send me to the local shops to buy or collect some item of food. She would give me our ration book and the shopkeeper would take a coupon as well as money. To me the ration book was part of shopping, so I couldn't quite understand why the coupons weren't needed when the government ended rationing. In the war and postwar years, rationing in the UK was introduced by the government to deal with the sometimes-extreme shortages of imported food. I remember that we only rarely had jam on bread and never on scones. Jam was a Sunday treat.

Scarcity was a part of life in 20th century Britain. For example, in 1916 it was illegal to eat more than two courses for lunch in a public place. People were to eat slowly and only when they were hungry. They were told not to feed stray dogs. It was perhaps unsurprising therefore that the early 20th century, the British economist Lionel Robbins (1898–1984)[13] widened the scope of economics, as then defined, to include any human activity concerned with the allocation of scarce resources among unlimited uses. He famously said, *'Humans want what they can't have'*. He clashed with perhaps the most famous and influential economist of them all John Maynard Keynes (1883–1946). Keynes thought that periods of unemployment were inevitable when the economy was contracting because the goods produced might not be needed (what economists call demand). If the aim was full employment, then the government had to intervene to give people purchasing power through government spending.

These assumptions as laws, ideas and purposes of economics are just a few of many—economic theory is replete with them. So how should we think of them? As I said earlier, we should think of them all (whether laws, theories, or purposes) as assumptions expressed as rules. They should be used very carefully and in a suitable context because as we said at the start through the words of the Buddha *'We are what we think...with our thoughts we make the world'*.

IS THERE A FREE LUNCH?

Everything we make, manufacture, or build, has a cost. Something gained freely means someone or something else pays for it in some way. Every criminal theft, apparently free to the robber, but a cost to someone—the victim and perhaps an insurance company. Every act of pollution is a cost to our environment. Every government benefit is paid for from government income obtained from taxes or borrowing. Although the person receiving unemployment benefit may not understand where the money comes from and as we shall see where government money comes from is hotly disputed. Spending is a cost for the buyer and income for the seller. Costs tend to rise as income rises. It seems to be common sense that there is no such thing as a free lunch—everything has to be paid for. However, as we shall see later in this chapter and in Chapter 5 when we do look at where money comes from (economists call it the money supply) we will see that government's that issue their own money (so-called *fiat money*) can have a free lunch. But it isn't a lunch without constraints. The most important obstacle is the need to recognize the availability of real resources and hence to contain inflation and maintain low unemployment. These are two of the most common risks to economic health.

THE BASIC FOUR CAUSES OF INFLATION

First, if a government spends more in an economy with full employment, then we get inflation. Economists call it demand-pull inflation. This is simply because the resources required to meet the extra demand will not be available. There is too much money chasing too few goods. This condition is rare since full employment at a living wage is uncommon.

Second, inflation results when internal supplies are short. This is called cost-push inflation. It is particularly severe when consumers cannot buy alternatives or wait until prices fall. If this happens then supply chains and markets need to be looked at.

Third, inflation results from shortages of external supplies—another form of cost-push inflation that is particularly hard to deal with. Intentional reductions in supplies of external energy by suppliers such as Russia during the war in Ukraine starting in 2022 are an example. Governments find these situations difficult to manage—in the long term they have little choice but to stockpile oil where possible but at the same time to reduce their reliance on those unreliable suppliers. Government price caps on companies with energy monopolies have pros and cons. Whilst they reduce costs to purchasers, they do not tackle the underlying causes of inflation, they may reduce incentives, are bureaucratic and wasteful with a chance of a growing 'black market'.

Fourthly, inflation can result from poor governmental policies. Whatever one's views about Brexit the economic consequences are, at least in the short term, inevitably, inflationary through increased tariffs and bureaucracy. Giving people more purchasing power (to buy things) by lowering taxes can also be inflationary when supplies aren't adequate.

TYPES OF UNEMPLOYMENT

There are at least three types of unemployment, frictional, structural, and seasonal. Frictional unemployment is where workers leave one job for a better one. There is a necessarily a constant turn-over of staff in any reasonably healthy economy. Workers who lose their jobs because of inadequate skills or income are the victims of structural unemployment. This is especially a problem where new technologies take over from older more traditional industries as occurred during the industrial revolution and is occurring now through increasing use of digital and IT communications.

Seasonal unemployment occurs when companies lay off workers because there is insufficient work for them to do. More generally when consumer demand is low then companies have to lay-off workers. Unemployment can then become cyclical. With fewer workers, demand weakens even more. A cycle of decline can spiral out of control. Governments have to intervene by expanding the money supply, for example by lowering interest rates to make loans cheaper, encouraging spending and boosting market confidence. They may use fiscal policies to increase or decrease spending and taxes.

The labour market is different from the buying and selling of goods. This is chiefly because employers hire the time of their workers and inevitably the productivity level of those workers varies. Productivity is not simply about how hard people work or getting more things done every day. Rather it is about getting the important things done consistently. Productivity depends on many factors—personal and collective motives, good leadership, teamwork, personality, natural talent, training or education, environment, support from others, time management, and even luck. Physical exercise, healthy eating, and sufficient sleep are all important too. Governments and employers can improve productivity by providing and maintaining good infrastructure, such as, education, training, transport, safe and healthy working conditions.

But why do government pay people not to work? Essential though social benefit schemes are, why do governments not pay people to do something useful? Why keep people idle in unemployment and subsidize their wages when governments could employ them to work such as rebuilding and maintaining infrastructure. An added advantage is avoiding long term (over 1 year) unemployment after which it is difficult for people to get back into work.

A jobs guarantee scheme[14] cannot cause a wage spiral and cost-push inflation, nor can it create demand pull inflation because it is limited to those currently unemployed. Currently governments stabilize prices for the advantaged only by forcing involuntary unemployment onto the disadvantaged. When workers are made redundant, they go home and do nothing. Extended unemployment causes skills and experience to decline, damages social interactions, limits options, decreases savings and increases debt. Businesses prefer to hire workers with skills and experience so unemployment may become permanent. A jobs guarantee scheme creates transition work until a more attractive job turns up.

SHOULD WE JUST GO SHOPPING?

It is perhaps not widely appreciated by non-economists that money and wealth are different. Money enables us to buy things and is an instrument of exchange. The wealth of a person or firm is the excess of assets over liabilities. The distinction becomes clearer if we think about a castaway on a desert island. She would not be any richer if she found a gold mine or a case full of bank notes on her shoreline. The reason is that she couldn't spend it on her island. In her case a fishing net might be a better source of wealth. In general, the relationship between wealth and money is not straightforward. People put different values on a good or service. A good meal once a day maybe valued more by someone in poverty than a wealthy man who eats well regularly. Their wages will depend on productivity because companies will hire workers if their rate of increase in productivity is higher than the rate of increases in wages. Trade increases wealth when a good or a service is transferred to someone else who values it more. For example, if a company makes batteries which they sell at a profit, then the wealth is shared by everyone involved including the company, the salesmen, the transport company, and any others who are in the supply chain. Profits are not just pay-outs as dividend to shareholders or wages to staff, profits are also vital to new investment in both a company and in R&D.

If the value of a product depends on its utility and not on the labour used, then aiming for full employment makes no sense. By this view, a useful product must benefit the consumer. Its value is independent of the effort to make it. For example, elite professional sportsmen do not get paid more because they take more time and effort, rather they are paid very large wages because they perform better and win more often.

Jon Alexander[15] calls this the consumer story—summed up as '*Just go shopping*'. He says that the consumer story goes like this: '*each of us is out for ourselves, and that is the way it should be. We are individuals, narrowly defined and independent of one another; the 'self' might extend as far as our immediate family, but no further. Human nature is lazy, greedy, and selfish, but can be overcome if we set our minds to it and work hard*'. He sees the consumer story as leading us into a spiral of decline. It assumes that we are narrow individuals. It tells us that we are not the ones to solve the problems of the world. We are limited to choosing between options that others provide. All we have to do is buy better stuff.

Here is another example of economics being, at once, both simple and complex—especially for the non-specialist. We have to choose between a bewildering array of multiple purposes and assumptions about how financial systems work. They interact in ways that can be foreseen and in many ways that cannot.

We cannot be left just to do the shopping, important though that is. Economics is too important to be left to economists. Economics affects us all in our day to day lives. It is determining how we tackle the big existential challenges of the 21st century of climate change, pollution, and inequality as well as the ways in which we relate to each other across regional, national boundaries. It is slowly killing the golden goose. So, what have we learned so far?

SEVEN LEARNING POINTS

We have glimpsed at the general role of assumptions in our lives both generally and in economics. Now I want to share and discuss the first learning points about economics that I have personally found revealing as a non-specialist. I do this in the hope that it might help others who are also non-specialists to understand a little more about how economic and financial systems work and how they affect all our lives in quite basic important ways.

My learning point Number 1 is that there are three surreptitious assumptions at the heart of capitalist economics. They are surreptitious because although they are not purposely disguised or properly referred to, they are not given major prominence and so one gets a sense of them being smuggled in. I will start by stating them very briefly and gradually expanding them later in the book. In overview, they concern the role of scarcity in economic theory, economics as a science *vis a vis* the way it is practiced and the idea of the economic man and utility that tend, on the face of it, to associate capitalism with rewarding greed.

Before beginning to amplify these first three assumptions, I want also to share *my learning point Number 2. There appear to me to be (at least) three important and wide misunderstandings amongst economically influential non-specialists such as politicians.* These are how we justify the case for a capitalist economy, notions of equilibrium, simplicity, and complexity and finally the role of fiat money and debt and the consequences for fiscal and monetary policy and money supply. How do politicians justify judgements about policies around the costs of government borrowing that are so often unsuccessful. At the root of these misjudgments are misunderstandings that have created deep divisions within macroeconomics between mainstream thought and Modern Monetary Theory (MMT). For example, the Bank of England[21] reported in 2014 that the reality of how money is created today differs from the description found in some economics textbooks. Most commonly the view held is that banks simply receive deposits when households save and then lend them out. In fact, the deposits increase bank reserves. The banks then keep the reserves at a level decided by the central bank and lends out the rest. The net result of bank lending is that the supply of money into the economy increases. In normal times, the central bank does not fix the amount of money in circulation, nor is central bank money 'multiplied up' into more loans and deposits.

My learning point Number 3 derives from MMT. It is that Government spending creates the money to pay tax and fund government borrowing. Money is not needed to fund the deficit. The deficit makes the money to pay for itself. Government borrowing is the redepositing of the excess of government spending over taxes paid back to the government[22]. Learning points 3 and 4 are critical to the financing of climate change. We will return to them in much more detail in Chapters 4 and 5.

But now, let's look at learning points 1, 2—the three surreptitious assumptions and the three wide misunderstandings.

The first assumption can be restated in a slightly more detailed form as saying that the definitions of economics, variously referring to the 'science of making choices under scarcity,' are partial[16].

Economics is much more than these definitions imply. First it requires value judgements—something many economists deny whilst others distinguish positive from normative economics. Positive economics, they argue is about describing 'what is there' whilst normative economics says whether a situation is good or bad. As Carney[17] recently observed society has, unfortunately, come to embody Oscar Wilde's old aphorism: *'knowing the price of everything but the value of nothing'*. These values include all forms of capital—natural, human, and social. Air and water are *not scarce* but are inherently valuable for life on earth. So, economics must be more than the study of scarcity.

These assumptions are surreptitious because they seem innocuous. Scarcity means there is a limited amount of something and that leads to us having to make choices. It is plain commonsense that if we choose to build a bridge over a river rather than a hospital, then we satisfy one need (being able to cross the river) but give up another (having a new hospital). The cost of not having the hospital is an 'opportunity cost'—an idea we highlighted earlier. Economics then becomes a discipline that looks at how scarce resources are used to satisfy needs—but more than that, how some countries are rich, and some are poor. Scarcity is central to Adam Smith's law of supply and demand. This also seems innocuous. Smith's law states that, all other things being equal in a competitive market, the unit price for a specific good, or asset, will vary. But it will eventually settle down when the quantity demanded equals the quantity supplied and there is a kind of static economic equilibrium. We'll see later in this chapter how the simple idea of a kind of static equilibrium is inadequate—the equilibrium is actually dynamic and may even occur in, what some call, a 'far from equilibrium' complex state.

There is no explicit mention of scarcity in Smith's assumptions—but it is implicit. Scarcity is the gap between limited resources and limitless wants. It is the idea that there is never enough of something to satisfy all conceivable human wants. It means making a trade-off to obtain more of what is wanted. Lionel Robbins[12] made scarcity explicit when he defined economics as the *'science which studies human behavior as a relationship between ends and **scarce** means which have alternative uses'.*

So now we can see the folly of economic theory when economists assume that people make decisions about how to allocate resources efficiently according to scarcity. That folly is the disregard of some of the things that seem to be not scarce but are most valuable to us—the cleanliness of the air we breathe, the potability of the water we drink, the biodiversity of the natural world and the prevention of degrading our environment by plastic waste and other forms of pollution including the emission of greenhouse gases. Economists tend to call these factors 'externalities'. In other words, they are outside the economic system—but that is not good enough—the system has to be more inclusive.

In certain circumstances, free natural resources can become scarce—lack of water in a drought or lack of warmth in an extremely cold environment. Water falls freely from the sky as rainfall but there are costs in making it potable and delivering it to our taps. If we want something badly enough, we may give up anything or everything for it. One of Shakespeare's most recognizable quotes is Richard III crying out *'A horse, a horse! My kingdom for a horse'.* The idea of a king wanting a horse so badly that he would give up his whole kingdom is a dramatic overstatement. But when something is wanted that badly, then one will give up anything to get it. Money and time are often scarce resources. Too little of one, or both. An unemployed person may have time, but no money. A high-flying businessman may be financially well-off but pressured into eat snack lunches and sleeping only four hours a night. He has plenty of funds but is short of time in a way that harms his mental and physical well-being.

We deal with scarcity by increasing supply or reducing wants or demands. Increasing supply has limits. There may not be sufficient available land, time, labour, or other resources. Poor air quality can limit supply through high rates of disease and death. Remedies have to be paid by the citizens in one way or another. In reality, breathing our free air is not free.

We can conclude from all this that scarcity is a sufficient condition for economic decision making but it is not necessary. There is more to economics than the study of scarcity because there are other valuable factors (externalities) that would better be brought into the economic system—such as biodiversity.

Equilibrium in economics is widely misunderstood. The simple notion of supply and demand in 'balance' in an economy of static equilibrium is held by many including politicians. It implicitly assumes that all is in a stationary or in a steady state with perfect coordination. Now we are realizing that these kinds of assumption are no longer adequate. They derive from traditional science that emphasizes stability, order, and equilibrium in closed linear systems where small inputs lead to small outputs. They have served us well in engineering well behaved physical systems such as buildings, bridges, engines, and aeroplanes. But as our thinking has developed to consider much wider systems that must be sustainable in nature and the environment, we are beginning to observe complexity. Modern theories of complexity recognize that we live with accelerating change in open systems. It's a kind of change that leads to disorder, instability, diversity, and disequilibrium. Open systems exchange energy, information, and material with their environment. We are now starting to describe information as organized patterns of energy[18]. Fluctuations and instabilities are intrinsic, so that economies do not converge to stable stationary situations of steady uninterrupted economic growth. Complex systems are modeled by partitioning its state space, or sets of all possible known states of affairs, into zones or phases. Processes may switch from one such zone to another with a different structure. A single fluctuation may result in a bifurcation, or forking at a singular point that can't be predicted. Day[18] notes four properties of economic behavior as (a) irregular fluctuations of economic data, (b) structural change, (c) technical innovations and (d) structural breakdown.

These are the reasons why many people say we are living in a complex age of information. Relationships are non-linear and small inputs can lead to very large outputs. With this new awareness we have to think differently about the meeting of supply and demand to determine prices. At the very least the balance is a dynamic, ever-changing state—a kind of 'dynamic togetherness'. As I said earlier some economists are beginning to call it as a 'far from equilibrium' state[18] meaning far from static equilibrium. Static equilibrium structures exist as isolated systems but systems that can only exist in symbiosis with their surroundings were named as dissipative by Ilya Prigogine a Belgian Nobel Prize winning chemist. Dissipative systems are of two types. Either close to equilibrium where their order tends to be destroyed or far from equilibrium where order can be maintained, and new structures formed through a continuous exchange of energy with the outside world. The forming of dissipative systems demonstrates that it is possible to create order from disorder. Prigogine and his collaborators say that what we observe in reality as different hierarchical levels of organisation directly relate to a succession of instabilities. This dynamic equilibrium means that state of a system is constantly changing with time through instabilities and an external input of energy or matter. Consequently, some economists see the economy, like life itself, as a self-organizing evolutionary process that produces order out of disorder and chaos.

The state of the economic system is commonly, but inadequately, described in terms of Gross Domestic Product (GDP), inflation, and employment (and many more). The state is moving around a zone, stage, phase, or region of values with models of behavior particular to those phases. Evolution occurs through the switching between the phases caused by a large perturbation such as a shock or an inherent instability. Crises carry the current system to a far from equilibrium state which can precipitate a very rapid restructuring of the economy and society[17].

The second surreptitious assumption concerns economics as a science *vis a vis* the way it is practiced. Economics does rely on social science but ultimately it is practical. Most economists regard economics as a science. As an engineer, I see this as a category mistake. First, I recognize that there is certainly a social science of economics, and some theoretical economists build mathematical models. However as soon as the science is used to make practical decisions it becomes a form of engineering. By engineering I don't mean something-to-do-with-engines, but I do mean engineering in the general sense of skillfully or artfully contriving to bring something about. As noted earlier many economists recognize that economics is normative, in other words it is not a matter of true or false but of good or bad and of how things *ought* to be rather than what *is* the case. It involves value judgements just as engineering does. But practical economics goes beyond being a normative science—just as engineering does.

You could be forgiven for thinking that the difference between science and engineering is trivial. Surely engineering is simply the 'appliance of science'. This is how the media often portray engineering as technology. But this widely held view is misleading and can be unfortunate. When we look at why things go wrong—for example, when human error leads to disaster as it did at Chernobyl[19], we find a much more complex situation. *Purpose* is an all-important idea. When science is applied in practice it is far from being a pursuit of truth. As I have said all theory is context sensitive, and to some degree fit for the purpose we have set ourselves. Satisfying that purpose might involve conflicting goals and the need for trade-offs. In economics the rules for determining the value of free but highly valuable natural and human resources must be different. They require reconciliation. For example, if we aim for full employment rather than consumption, the economic rules we follow may be quite different. Government policies that keep the unemployed idle by subsidizing their wages is a choice. Governments could employ them to work in areas such as rebuilding and maintaining infrastructure with the added advantage of avoiding long-term (over 1 year) unemployment which makes it difficult for people to get back into work[13].

Most economists agree that economic growth depends on new technology even though some of it may be disruptive. This is particularly so in developing countries. But rooting out corruption could be an even greater need as we shall see in Chapter 7. The boundaries are thin. In the USA in the late 19th century people such as Vanderbilt, Carnegie and Rockefeller built enormous businesses through 'cutthroat' capitalism. They created gigantic monopolies by intentional restrictions and price rises. They were called 'robber barons' because of their ruthlessness in business. Yet they did later make some amends by giving hundreds of millions of dollars to good causes. A whole network of Carnegie libraries was endowed across the English-speaking world—indeed I used one as a child in my hometown. The 17th century slave trader Edward Colston was a philanthropist who endowed schools and public institutions in Bristol, UK. But in June 2020 protests against the slave trade and petitions for name changes resulted in his statue being toppled and pushed into Bristol Harbour during protests in support of the 'Black Lives Matter' campaign.

Just as this business ruthlessness was going on there was another instinct. The impulse to do productive work, with good workmanship, improved methods, ideas, and machines. The impulse to improve the human condition[20].

Economics ought to be about human flourishing—what it takes to live well—to be happy and fulfilled. The third surreptitious assumption is important here—the idea of the rational economic person (*homo œconomicus*) as an idealized person who acts rationally, with perfect knowledge, to maximize personal satisfaction which is called utility. We will discuss this in much more detail later in the chapter.

The point for now is that, as currently conceived, the purpose of economics isn't formed around human flourishing, it is formed around self-gratification. Controlling inflation and providing employment are part of economics, but they are not the whole—if we see its purpose as helping humans to flourish. A minority of economists have held this view so far, but in 2000 Sir John Templeton, investor, and philanthropist, wrote that *'freedom fosters the kind of constructive competition that makes progress possible. When the creativity, ingenuity, and competitive efforts of individuals are set free, the results can be progress and prosperity beyond anything ever before imagined'*.[22] Sir John was a firm adherent to the philosophy of the classical liberal tradition, which emphasizes individual freedom of choice, voluntary association, entrepreneurship, and markets in the ordering of social life. But the age-old questions still need to be addressed such as whether societies work better with free markets and competition or by central command or by cooperation. Nevertheless, a focus on the purpose of economics as promoting human flourishing would be clearer and 'bigger' in the sense that it would lift our horizons above consumerism. It would enable us to address questions as to why we are here and how we can survive for a better life.

Jon Alexander[23] is on the right track when he calls for us to adopt the citizen story of collaboration and cooperation. Within that vision economics requires us to ask questions such as, should we tax carbon at a level that will influence the decisions that people take? Should governments issue carbon trading permits? Can we solve problems by balancing costs and benefits? Governments can redistribute resources by taxing the rich to give to the poor, but the counter argument is that this upsets efficiency and results in slower growth. Nevertheless, bestselling economist writer Thomas Piketty[24] proposed a global tax on the rich. Inequality results, at least partly, from the choices made by societies and we certainly should be promoting engineering technologies that reduce inequality and encourage citizenship.

But practical life is difficult. Thomas Carlyle[25] accused economics of being a 'dismal science' because he agreed with Malthus that population will always grow faster than food production—doomed to unending poverty. But equally, he might have described it as dismal because it is a set of theories that can easily be found wanting and for which viable alternatives are not straightforward. In Chapter 3 we will look at how economists assume that markets are efficient, and people are rational. Gross Domestic Product (GDP) is one of the main tools in their toolkit. It is constantly in the news as the media report to us whether it is growing or declining. But almost all economists agree that GDP is a poor measure of economic performance. It omits relevant factors such as the contribution of housework and bringing up children all of which are important to the health of the economy.

The regular occurrence of crises is a worry. An economic crisis is when the economy contracts. If it lasts for more than 6 months—measured as a fall in GDP—it is called a recession. Recessions undermine future social mobility, lifestyle, mood, mental and physical health, social stability, and insecurity. Financial crises are slightly different. They are when we get speculative bubbles, bank runs, pandemics, falls in house prices, trade wars, changes in oil prices, wars, famines, and even social unrest. Financial crises lead to lower wages and business closures, unemployment, greater inequality and, inevitably, economic crises. Economists say a 'Minsky' moment (after the economists Hyman Minsky, 1919–1996) is reached when people quite suddenly and collectively realise that their shares are worth a lot less than they thought. In that case speculation has reached an unsustainable extreme—a tipping point. For example, although, at the time of writing, the precise causes of the collapse of the Silicon Valley Bank (SVB) bank in 2023 are unclear. The bank was set up to serve the most dynamic and innovative sector of the global economy, i.e., 'tech industries'. It had placed large deposits in low-yield—but safe—government bonds. Then the Federal Reserve raised interest rates and the value of these bonds declined, setting off a bank run as depositors scrambled to withdraw their money.

Then came the unintended consequences on other larger banks like Credit Suisse.

A crisis due to a completely unexpected event such as the Covid-19 pandemic is called a 'black swan' event because most swans are white. But despite all the harm crises can bring some good. Bad businesses are likely to go bust. Good businesses are likely to survive and improve. New businesses may be created, and excessive hype moderated. These events are called 'creative destruction' by some economists. Joseph Schumpeter (1883–1950) first used the term to describe how the old is constantly being replaced by the new. For Schumpeter, the entrepreneur was the keystone of capitalism—the source of the vital innovation that drives a capitalist economy. He said, *'Innovation is the market introduction of a technical or organisational novelty, not just its invention'.*[26]. Boom and bust derive from waves of innovation. New technologies displace old ones. Equilibrium is the economy in a freeze frame of constant movement.

Perhaps one of the biggest crises of the past was the Great Depression of the 1930s that started in the USA after a major fall in stock prices. How did USA, the world's richest economy at that time, succumb? Keynes, who is still very influential said that although classical economics was about scarcity, it really should be about how to make use of existing resources. The crisis occurred, he thought, because the connection was broken between what people wanted and what the economy was making. If people save rather than spend then it follows that those levels of investment and economic activity reduce. Keynes put his finger on the pulse when he saw that the spur to economic growth was the optimism of the people involved. Consequently, he was much keener on fiscal policy than monetary policy—macro policy options that we will look at in Chapter 3. He saw unemployment as being caused by a lack of demand for the things that require workers to make them. So, he thought that government interventions are needed to avoid recessions. He wanted savings to be invested because that would create economic activity and growth. In other words, government spending is a means of stabilising the economy rather controlling the money supply (see Chapter 5). Unfortunately, before the financial crisis of 2008 many economists thought that government should reduce company taxes and loosen restrictions to encourage the employment of more workers. They wanted to improve supply side economics. But the level of private debt became unmanageable due to unscrupulous and corrupt banking and insurance practices in subprime mortgages with inevitable consequences when the holders could not repay their loans. Other forms of crises are natural ones. For example, around 30–60% of the population of Europe in 14th century were killed by the plague called the Black Death, and from 1918 to 1919 Spanish Flu killed around 200,000 people in the UK and 17–100 million worldwide. Covid-19 killed 6–25 million in 2019–2021. The economic losses in every case were huge.

If economics is understood as promoting the human condition, then our goal is to enable human beings and the natural world to flourish. But to think that way we need to see economics as a form of engineering. Both engineering and economics depend on and use science, but they are ultimately practical—about doing rather than knowing, with theories that are value laden and good/bad, and uncertain and are not matters of truth or falsity.

ECONOMICS IS BOTH VERY SIMPLE AND VERY COMPLEX

So where does a little understanding of how economics has developed over history get us? As we have said, at one level economics is very simple but at another it is very complex.

At its simplest we all know that income must be greater than expenditure or else you get into debt. As Mr Micawber said in Charles Dickens's 1850 novel David Copperfield *'Annual income twenty pounds, annual expenditure nineteen, nineteen and six, result happiness. Annual income twenty pounds, annual expenditure twenty pounds naught and six, result misery'.* We all know that we must pay tax. Although the many and various forms of tax may elude us, we are familiar with paying income tax and VAT. We all know that interest is the amount a lender charges a borrower as a percentage of the amount loaned. We expect to get interest on any savings we make with a bank or building society. We expect to pay interest on any loans or mortgages we borrow to buy a car, house, or apartment. Perhaps less well known is that demand and supply of goods and services will be in equilibrium when they are equal. But as we saw earlier the nature of that equilibrium is complex. A price of something above that equilibrium shows that there is a surplus of supply above that needed. A price below that equilibrium shows that there is an excess of demand or shortage of supply. The risk is that the very idea of price equilibrium suggests that the system is static when it most certainly isn't. That suggests that there are hidden complexities that most of us who are not economists do not think about. We are aware that economic forecasting is to be taken with 'a pinch of salt'. Economist Ezra Solomon once quipped that "*the only function of economic forecasting is to make astrology look respectable*"[27]. We are aware of the disagreements when politicians and economists debate very publicly which economic policies should be adopted. But most of us have only a hazy understanding unless we have taken the trouble to educate ourselves in matters economic. We leave these things to the experts. This would be acceptable were it not for the fact that their decisions have very significant and serious effects on our standard of living.

Economics is complex—but not *mainly* because the principles are difficult. There are multiple reasons for its complexity. Some, but not all, go very deep because they depend on our behavior as intentional human beings who have different motivations, purposes and levels of knowledge and understanding. Complexities arise from, for example, the considerable delays between cause and effect, the multiple ways in which financial entities interact, the non-linearity of mathematical relationships in economic models, a great deal of uncertainty, emergent properties (that arise from the interactions between the parts), conflicting requirements and the consequent difficulties in making predictions. These complexities manifest themselves in many ways. For example, financial entities can be bankrupt before they realise. Prices as expressions of demand can be in a state of unstable equilibrium with supplies of goods—even far away from equilibrium but all the time changing dynamically. The business cycle of growth and decline (boom and bust) is an expression of natural complex change. Over time we may have learned to manage its worst effect better, but we are not able to banish it completely. Financial crises still occur, for example in 2008 when there was too much private credit on unsecured loans when compared to gross domestic product (GDP) and in 2022 after the shocks of Covid-19 pandemic and the invasion of Ukraine by Russia and the step increase in the price of energy.

Engineering and technology have a profound impact. Economic development depends on the predictability and uncertainty surrounding engineering creativity. In the 1960s few of us foresaw the massive development of the transducer into the printed microelectronic chip and its impact on information technology (IT) in the form of computers, the internet and mobile (cell) phones. Even the invention and manufacture of the humble pencil[28] resulted from an entrepreneurial spark of creativity. That creativity cannot easily be foreseen or planned. The result is an economic dynamism. The freedom to enter and exit markets and the competition it engenders enables entrepreneurs devise goods and services that improve the lives of consumers. The profits can be sizable. Amazon's commercial success derives from using technology to simplify the acquisition of a wide range of retail goods. Major international companies compete to offer ever-better products that simplify household chores.

THREE EXAMPLES OF COMPLEXITY FROM SIMPLICITY

Let's consider three examples of complexity from simplicity, money, interest, and tax.

Money

In Chapter 5 we will look at money in more detail. Money is simple in the sense we use it every day. We know we must try to make more money than we spend otherwise we get into debt. We know can borrow money to buy things we can't afford immediately—for example, taking out a mortgage to buy a house or a loan to buy a car. We realise we must pay interest on those loans, and we benefit if we are able to save money in a bank, building society or other account and get some interest. We may even invest in businesses or other financial institutions by buying shares or bonds whether public or government (gilts in the UK). We are familiar with money as coins and paper notes. Perhaps we are less familiar with money that is not physically available but exits as entries in the ledgers of banks and financial institutions. We may have some realization of this as increasingly we rely on electronic banking and pay for goods using a credit or debit card. We have some understanding of money as a medium of exchange, a store of value and a unit of accounting. Economists have several designations of money and there are confusingly different ways this is done. This when money starts to become more complicated. Economists distinguish between narrow money and broad money. Cash is only one form of money. Money comes in five forms according to liquidity (ease of access or convertibility into cash without price change) but the exact classifications vary between countries. M0 is coins, paper notes, and central bank reserves. M1 is M0 plus readily available bank deposits and travelers' cheques.

M0 and M1 are called narrow money. Sometimes the term near money is used to describe assets that are not cash, such as bonds, time deposits and shares, which are very liquid and hence easily convertible into cash. M2 money includes M1 plus short-term savings deposits, money market funds, certificates of deposit, and other time deposits. M3 money includes M1 and M2 plus deposits with an agreed maturity of up to two years, deposits redeemable at notice of up to three months and repurchase agreements, money market fund shares/units and debt securities up to two years. Sometimes the term M4 money is used to describe M3 plus other deposits. M2, M3 and M4 are called broad money but vary according to local practice.

If we didn't have money, then we would have to barter and directly exchange one good or service for another—very cumbersome. To be of any use money must hold its value over time and we must be able to use it as a common measure of exchange. We will look at the historical story of how that has evolved in Chapter 5.

We have to admit that personally we are not always rational about money. We don't always make or follow a budget or make regular savings. We rarely draw up a financial plan. Some small businesses don't make a business case—they just evolve as work comes in. We may feel guilty, fearful, even shame about our spending habits and the state of our finances. We may feel envious of those better off than we are. We may avoid thinking about it or doing the things we should do. For example, we may have mixed emotions about charitable giving. Shame and avoidance can cause a vicious circle.

At the second level of complexity governments have policies to manage money—as we have seen there is a school of economic thought called monetarism. It maintains that the money supply, or the total amount of money in an economy determines economic growth. Money supply is the amount of money circulating in an economy. Governments can control the rate at which the money supply grows. As the money supply increases, people demand more, and companies produce more, thus creating new jobs. However, this is only temporary because continuing to increase the money supply can cause inflation when demand outstrips supply and prices rise. Governments around the world have handed setting policy on money to central banks such as the Bank of England or the Federal Reserve in the USA. As a consequence, countries normally have two sets of economic policies, fiscal and monetary. Fiscal policy is the way governments spend money and set taxes. Monetary policy is the way the supply of money is controlled. We will delve more into the complexities of monetary policy in Chapter 5.

Interest

An interest rate is the amount that a lender charges a borrower. It is measured as a percentage of the amount of the loan. It is typically quoted as an annual rate or APR (annual percentage rate). At the simplest level, people who save money get interest for leaving their spending of that money to a later date. Lenders get interest as a reward for not being certain that their money will be repaid as agreed. More generally interest is a charge to a borrower for the use of an asset which can include money, goods, and property. It is essentially the cost of money because higher interest rates make borrowing more expensive. As well as individuals taking out mortgages to buy houses or loans to buy cars, so do businesses borrow to fund big projects or buy land, buildings, or machinery. The loans are repaid either as a lump sum or in installments over an agreed period. The interest rate is known as the cost of debt (for the borrower) and the rate of return (for the lender).

But who sets the rate of interest? Some economists think that there is a natural interest rate set by a market, which satisfies both savers and lenders when the supply of money meets the demand for it. However nowadays governments or central banks intervene in that market mechanism and impose a basic interest rate called the bank rate. That is the rate that a country's central bank charges other banks to borrow from them. The interest rates that other lenders, such as high street banks, charge depends on the bank rate plus other factors. The chief one is the risk of the loan not being repaid. The greater that risk and the longer the period of the loan then the higher the interest rate.

Interest gets more complex when it is used to control the economy. Central banks have one principal lever—the raising or lowering of the bank rate—though they can also change the reserves that high street banks are required to keep back. Raising the reserve requirements restricts the capacity for banks to lend and slows economic activity. Lowering the bank rate helps expand the economy by lowering the costs of borrowing. When that happens, we have a period of 'easy money'. Alternatively raising the bank rate helps to squeeze the economy—for example, when inflation is too high. Economists often disagree about the unintended consequences of changing interest rates. For example, does easy money impede or encourage people to allocate money to more productive resources. Easy money can encourage the taking of undue risks. For example, the banking crisis of 2008 was triggered by low interest rates creating a demand for credit. Bankers were innovative in designing new financial packages with risky short term and adjustable interest rate mortgages. There was a consequent growth in complex mortgage securities (interchangeable assets) as investors were seeking better returns on their loans. The bubble burst when financial institutions realised that they were holding trillions of dollars of near worthless investments and homeowners owed more than their houses were worth. Companies that specialized in subprime mortgages went bankrupt (more on this in Chapter 5).

Interest rates when used to control inflation can also have unintended consequences. The orthodoxy is that higher interest rates raise the cost of borrowing so that people and businesses borrow less and spend less. Central banks can change the reserves that high street banks are required to hold. Raising the reserve requirements restricts the capacity for banks to lend, makes government slows economic activity. That results in lower levels of economic activity, higher government interest on debt repayments and dampens prices. It benefits savers but increases the cost of living for mortgage holders. In 2022, as I write this, the world hovers on the brink of a worldwide recession as the gross domestic product falls. Stagflation occurs when inflation is high and economic growth is low or negative in recession. The cost of living has spiked because of the shocks of the Covid-19 pandemic and the surge in energy prices consequent on the war in Ukraine.

The World Bank is predicting that stagflation could persist for several years with serious consequences for low and middle-income economies. Developing countries, where people go hungry every day, have record levels of external public debt. They have, in many cases, borrowed with variable interest rates. As global financing tightens (reduced money supply) and currencies depreciate, debts previously found only in low-income economies will spreading to middle-income countries. For example, Sri Lanka, a second world economy, has borrowed to the point of near bankruptcy from the International Monetary Fund in US$ and from other countries such as China and Iran. During the Covid-19 pandemic, tourism collapsed, and the Sri Lankan rupee declined against the US$ so that in April 2022 Sri Lanka has almost no foreign currency and cannot afford imports of food, fuel, and medicines.

The World Bank predicts that monetary policies across the world will not be able to accommodate sudden changes in supply to offset economic shocks—both in the immediate past and the unknown future. They say that there is a need to increase the rate of monetary growth to restore purchasing power (the amount of money a person or group has available to spend). Inequalities need to be reduced by promoting higher incomes, strengthening supply chains, small businesses, and the ways capital is allocated. They call for five actions:

- limiting the harm to people affected by the war in Ukraine by coordinating deliveries of emergency food, medical, and financial aid and by housing, supporting, and possibly relocating refugees.
- boosting the supplies of food and energy. Even mere announcements of future supplies would help reduce expectations of price inflation. All countries should reduce export and import restrictions.
- introducing ways to reduce or refinance debt to help the borrowers repay it.
- strengthening the preparedness for health shocks, including containing Covid-19, by expanding immunization.

- reducing dependencies on fossil fuels and accelerating the transition to low-carbon energy sources. Invest in electricity grids, cleaner energy sources, and greater energy efficiency. Climate friendly regulatory frameworks, with appropriate incentives and regulations of the uses of land.

Taxation

"Tis impossible to be sure of anything but Death and Taxes".[29] Tax is simple in that we all know we must pay it. Some of us resent it and some even try to avoid it legally and evade it illegally. Nevertheless, we are aware it is needed, in a civilized society, to fund activities that benefit everyone such as health, infrastructure, defence, and education. The main purpose of tax is usually understood to be to finance public spending. But the government can choose to use it for other purposes such as reducing inequalities and improving quality of life, efficiency, and promote economic growth. That is because tax is not actually needed to fund government spending. You often hear politicians refer to taxpayers' money. They are wrong the money derives from government spending. But governments can't just spend money without limit because it would become worthless. Tax is what gives money its value because tax is the way governments can control the amount of money in the economy and hence help to control inflation. Income tax and VAT are the obvious taxes but there are many more we may be only vaguely aware of like Corporation tax, National Insurance, and inheritance tax.

Countries may also use their tax systems to compete in global markets but that, unfortunately increases the potential for evading and avoiding tax. Governments of a country may tax the profits from the products that are made there. The country where the company is headquartered will tax the company's total profits. Unfortunately, the country, where the final product is sold, has no taxing rights except for sales tax such as VAT so that multinational companies are not required to pay taxes to them. Some multinational firms reduce their tax liabilities by having a plant in a high-tax country selling its product to a sister entity in a low-tax country at below market price. This reduces the profits recorded in the high-tax jurisdiction and lowers the costs for the sister entity in the low-tax jurisdiction. The entity in the low-tax country then sells the final product to its distributors above market price to maximize revenue in the low-tax jurisdiction and book more profits at the lower tax rate.

Remedying this requires will require international cooperation. Government taxation policies affect savings, the supply of labour and level of investment in human capital to create jobs, invest and innovate.

The different ways taxes are raised are called tax instruments or tax structures. Many OECD countries have reformed their tax structures to encourage saving, investment, and entrepreneurship. Corporate tax reforms are intended to promote competition and avoid economic distortions due to poor drafting.

In the last 30 years the proportion of tax from personal income has reduced and top rates reduced[30]. The proportion of corporate income taxes has increased despite cuts in tax rates and decreased top marginal rate on dividends. Social security contributions have increased. Consumption taxes have declined, with greater general consumption taxes (mainly VAT). The share of property taxes and environment-related taxes has been fairly constant over time. Several countries have introduced tax incentives for investment in research and development.

Evidence[31] suggests that, in certain circumstances, recurrent taxes on immoveable property (e.g., buildings) is likely to be the best for economic growth. Then follows consumption taxes (and other property taxes), personal income taxes and corporate income taxes. Consequently, reforms of tax to promote growth emphasize taxes on residential property rather than on income. Shifting from corporate to consumption taxation would (a) increase share prices by increasing the after-tax present value of the firm, and (b) increase wealth and income inequality by lowering capital income tax. Flattening the income tax schedule could benefit GDP per capita by favouring entrepreneurship.

Taxes are said to be progressive when tax rates are higher for higher levels of income. Conversely regressive taxes decrease for higher levels of income, and they impose a greater tax burden on the poor than on the rich. Highly progressive income tax lowers macroeconomic growth because for some, it acts as a disincentive for getting a job that pays a higher wage. For instance, some retirees may choose jobs that pay a lower wage to ensure their tax rates don't change. In-work benefits can increase the income of low-income households, reduce inequality, and improve efficiency. This is because the gains in levels of employment outweigh (a) the reduced incentives to work as benefits are withdrawn and (b) on improving human capital as the returns from up-skilling are reduced and (c) the costs of the tax increases needed to finance the in-work benefits.

Tax distortions occur when people do things for tax reasons that otherwise make no economic sense—choices are motivated by tax and not just economic costs and benefits. Whilst making sense for individuals these actions are a real economic waste for countries. Corporate taxes affect levels of investment so that lower rates reduce distortions and attracts investments both domestic and foreign. Predictability in tax policy is important to companies and may lead to higher investment and enhanced growth. Lowering corporate tax rates can lead to larger gains in productivity dynamic and profitable firms.

Almost everything we think about and the actions we take depend on models. Models are not just toys but are any way we represent our understanding of the world we live in—including science. Models are cool—acceptable, admirable, useful, and fundamental. But we must be aware that models' may well express assumptions that are hidden from clear view. Either we do not realise that they are being made or we neglect them and allow the models they support to be used out of the context in which they apply—with potentially harmful consequences.

We cannot understand the models we rely on today without understanding where they came from. History tells us that unintended consequences abound. Shocks come in many forms. We cannot decide what models to use in the future without understanding the models we have had in the past and how they have evolved into the models we have now. We must always be scanning for potential unintended consequences and shocks. This is an important lesson of history.

Feudalism was a way of structuring society around reciprocal legal and military obligations between lords, vassals, and the land. Wealth came from agriculture, not through markets but through labour of the serfs to the lords. Serfs paid rents, provided food, worked the land owned by the lord, and helped defend the land from intruders, violence, and warfare. Feudalism eventually developed into mercantilism—the first stages of capitalism. Although banking in its various forms has been around since 2,000 BC, modern forms of banking developed in Italy in the Renaissance with merchant banks based on the Lombard plains cereal crops. Marginalism came from needing to understand how markets operated and prices fixed. The modern era started with Adam Smith and continued with Alfred Marshall and John Maynard Keynes. Later came thinkers such as Milton Friedman and monetarism where it is thought that variations in the money supply are major influences on national output and price levels. Policies should target the growth rate of the money supply.

I have highlighted three surreptitious assumptions at the heart of capitalist economics. They are the role of scarcity in economic theory, economics as a science *vis a vis* the way it is practiced and the idea of the economic man and utility. Three important and wide misunderstandings amongst economically influential non-specialists are how we justify the case for a capitalist economy, notions of equilibrium, simplicity, and complexity and the role of fiat money and debt. The simple notion of supply and demand in 'balance' in an economy of static equilibrium is held by many including politicians. It implicitly assumes that all is in a stationary or in a steady state with perfect coordination. Now we are realizing that these kinds of assumption are no longer adequate. Modern theories of complexity recognize that we live with accelerating change in open systems. It's a kind of change that leads to disorder, instability, diversity, and disequilibrium. Complex systems are modeled by partitioning its state space, or sets of all possible known states of affairs, into zones or phases. Processes may switch from one such zone to another with a different structure. A single fluctuation may result in a bifurcation, or forking at a singular point that can't be predicted. Now many people say we are living in a complex age of information. Relationships are non-linear and small inputs can lead to very large outputs.

As economics developed it became at once both simple and complex. Economics is simple in the Mr Micawber sense that income and expenditure need to balance. Complex not because the principles are difficult but because of multiple interactions that make it difficult to test hypotheses to establish theories. Although economists can apply the scientific method of critical discussion their ideas are not expressions of fundamental and absolute truth—rather they are models of often complex contextual situations. Context is as important as the model itself. It seems simple common sense that there is no such thing as a free lunch—everything has to be paid for. But government's that issue their own fiat money can have a free lunch—but not without constraints—such as the availability of real resources to contain inflation and maintain low unemployment. Examples of complexity are the four kinds of inflation, the many forms of money, interest rates and levels of taxation. If a government spends more in an economy with full employment, then we get demand-pull inflation. If internal supplies are short, we get cost-push inflation. If external supplies are short, we get cost-push inflation. Inflation is also by poor governmental policies. Frictional unemployment occurs when workers leave one job for a better one. Seasonal unemployment occurs when work depends on the seasons and companies will lay off workers when there is insufficient work for them to do. The consumer story assumes that we are narrow individuals. It tells us that we are not the ones to solve the problems of the world. We are limited to choosing between options that others provide. All we have to do is buy better stuff. We need to change to the citizen's story.

Money gets complex because opinions differ about how governments can control the rate at which the money supply grows. As the money supply increases, people demand more, and companies produce more, thus creating new jobs. However, this is only temporary because continuing to increase the money supply can cause inflation when demand outstrips supply and prices rise. Interest rates are complex because the amount that a lender charges a borrower affects so many aspects of economic life. From governments paying debt to individual paying mortgages, from easy credit money causing all sorts of unintended consequences to high rates increasing the costs of living. Taxation is complex because it can increase or lower the amount of money people have to spend.

Ultimately all we have is modelling—but we must be continually subjecting our models to tests in the real world and to constant criticism. Models of the economy are relatively simple—but because they are sensitive to context there is much room for judgement and interpretation. They are very difficult to test against reality because human behavior is infinitely variable. Consequently, economists agree on some fundamentals but disagree on interpretations of their models. That can be both puzzling and disturbing for the non-specialist. In a modern democracy where political decisions affect us all and economic policy making is central to our standard of living, we have a duty of care to understand the economic decisions that are made by our leaders before we give them our votes. It is clear to me in writing this book that misunderstandings are widespread. That behoves us to dig deeper into some of the complexities highlighted in this chapter. So next we will address one of the most fundamental assumptions of most economic models—that of the rational 'man'.

End Notes

1. Frigg, Roman and Stephan Hartmann, "Models in Science", The Stanford Encyclopaedia of Philosophy (Spring 2020 Edition). Edward N. Zalta (Ed.). See https://plato.stanford.edu/archives/spr2020/entries/models-science
2. Hagar, A. and C. Michael. 2022. "Quantum Computing", The Stanford Encyclopedia of Philosophy (Fall 2022 Edition). Edward N. Zalta and Uri Nodelman (Eds.). See https://plato.stanford.edu/archives/fall2022/entries/qt-quantcomp
3. Janiak, A. 2021. "Newton's Philosophy", The Stanford Encyclopaedia of Philosophy (Fall 2021 Edition). Edward N. Zalta (Ed.). See https://plato.stanford.edu/archives/fall2021/entries/newton-philosophy.

4. Blockley, D.I. 2010. Bridges. Oxford University Press, UK. Structural engineers apply Newton's Laws to find the forces in a long span suspension bridge such as Golden Gate in San Francisco.
5. Aristotle wrote this passage in his Politics Book1 [1252a] [1] and [1252b] [1]. https://www.perseus.tufts.edu/hopper/text?doc=Perseus%3Atext%3A1999.01.0058%3Abook%3D1%3Asection%3D1252a#:~:text=Every%20state%20is%20as%20we,they%20think%20to%20be%20good%EF%BC%89 .
6. Izdebski, A., T. Słoczyński, A. Bonnier, G. Koloch and K. Kouli. 2020. Landscape Change and Trade in Ancient Greece: Evidence from Pollen Data. The Economic Journal. 130: 632. Nov. 2020, pp. 2596–2618. https://doi.org/10.1093/ej/ueaa026
7. Wikipedia 2023. Guanzi. See https://en.wikipedia.org/wiki/Guanzi_(text)
8. Wikipedia 2022. Ibn Khaldun. See https://en.wikipedia.org/wiki/Ibn_Khaldun
9. Blockley, D.I. 2012. Engineering: A Very Short Introduction. Oxford University Press, UK.
10. Chancellor, E. 2022. The Price of Time. Penguin Books, UK.
11. Francois Quesnay (1694–1774), the marquis de Mirabeau (1715–1789) and Anne-Robert-Jacques Turgot (1727–1781) were the prominent physiocrats who produced the first well developed theories of economics that questioned mercantilism and immediately preceded classical economics. They saw human and animal muscle as the main source of power that produced agricultural surplus. Capitalist profit was the rent obtained by an owner of agricultural land. They praised natural styles of living and saw farming as the source of a nation's wealth. See https://en.wikipedia.org/wiki/Physiocracy
12. Wells, T. 2023. Sen's Capability Approach. Internet Encyclopaedia of Philosophy. See https://iep.utm.edu/sen-cap/
13. Lionel, R. 1935. The Nature and Significance of Economic Science, 2nd Ed. Macmillan, London.
14. Randall, W.L. 1997. Government as Employer of Last Resort: Full Employment Without Inflation. Levy Institute Working Paper No. 213 at https://www.levyinstitute.org/pubs/wp213.pdf . Also see overview at https://activistmmt.org/job-guarantee/
15. Alexander, J. 2022. Citizens: Why the Key to Fixing Everything is All of Us. Canbury, UK.
16. Definitions of scarcity include (https://www.economicsdiscussion.net/economics-2/definitions/top-4-definitions-of-economics-with-conclusion/14134) the 'study of scarcity and choice and its implications for the use of resources, production of goods and services, growth of production and welfare over time'. Others are the 'science of human behavior as a relationship between ends and scarce means which have alternative uses'; the 'study of wealth and of men'; 'that part of social welfare that can be brought directly or indirectly into relation with the measuring rod of money', and the 'science of making choices under conditions of scarcity'.
17. Carney, M. 2021. Value(s), William Collins, UK.
18. The analysis of the economy as a 'far from equilibrium' state starts by saying everything in the universe is an interplay between matter, and energy, and organisation as order (See Stonier Tom (1990). Information and the Internal Structure of the Universe. Springer-Verlag). Order is a non-random arrangement of parts, so randomness is the opposite of order. Unlike other models of dynamics, time becomes irreversible. Organisation is an expression and degree of order. Information is the raw material which when detected or observed is an event and which after processing may determine meaning. Information has the capacity to organise and maintain order just as energy has the

capacity to do work. Information can be a conjecture, plan, design, or economic policy. Information has to be processed to organise and maintain order. Entropy is a difficult concept but is effectively a measure of disorder. The relationship between information and entropy is complex and context dependent. As Alvin Toffler (see Toffler, A. 2017. Preface to Prigogine, I., Stenger, I. 2017. Order Out of Chaos (2nd Ed.) Verso) has written '*most of reality, instead of being orderly, stable, and equilibrial, is seething and bubbling with change, disorder, and process*'. There may be situations in which adding information increases entropy, and in others where it decreases. A system that tends to flow towards more ways in which it can be arranged and needs more information to describe it, such as an expanding gas, has a higher entropy. A system that becomes more ordered such as freezing water has a decreased level of entropy. A crystal lattice in a solid material has a highly ordered structure and low entropy. When heated the atoms or molecules in the lattice vibrate more vigorously. This increased energy and information leads to thermal ordering, where the lattice becomes more organized, and the atoms or molecules align themselves in a more regular pattern. As a result, the order of the system increases and entropy decreases. Energy is transformed by activity with some loss. More specifically the exergy of a system is the energy that is available to be used—the maximum useful work possible. It is a matter of ongoing debate whether life itself is order is created from disorder in a far from thermodynamic equilibrium [https://www.eoht.info/page/Far-from-equilibrium] since the term far from equilibrium is not well defined. Nevertheless, systems can self-organize, interact, and evolve by the dissipation of energy. Economic activity is the balancing of surpluses and deficits of goods by trading. Conventional economic theory ignores benefits or costs to third parties or externalities. For example, social systems and the environment that are not closed because they absorb and dissipate exergy from the sun to exist and evolve. See Ayres, R.U. 2022. The Economy as an 'Island of Order' far from Equilibrium. INSEAD Working Paper 2022/08/EPS/TOM at https://papers.ssrn.com/sol3/papers.cfm?abstract_id=4048996. See also Day R.H. 2009. The Divergent Dynamics of Economic Growth. Chapter 11. *Economics Far from Equilibrium.* Cambridge Books Online Cambridge University Press, See https://www.cambridge.org/core/books/divergent-dynamics-of-economic-growth/economics-far-from-equilibrium/9F101469F179BC99D8752F164969DE3E

19. Chernobyl Accident 2022. World Nuclear Association. https://world-nuclear.org/information-library/safety-and-security/safety-of-plants/chernobyl-accident.aspx
20. Blockley, D.I. 2020. Creativity, Problem Solving and Aesthetics in Engineering. Springer. On page 29. I promote a principle of ingenuity which says '*value, nurture and develop practical wisdom*'.
21. McLeay, M., A. Radia and R. Thomas. 2014. Money creation in the modern economy. Bank of England Quarterly Bulletin Q1. https://www.bankofengland.co.uk/-/media/boe/files/quarterly-bulletin/2014/money-creation-in-the-modern-economy.pdf\
22. Serazin, A., M. Bloom and T. Farquharson. 2022. The Purpose of Economics is Flourishing. Templeton World Charity Foundation at https://www.templetonworldcharity.org/blog/purpose-economics-flourishing
23. Alexander, J. 2022. Citizens: Why the Key to Fixing Everything is All of Us. Canbury, UK.
24. Piketty, T. 2017. Capital in the Twenty-First Century. Belknap Press, Harvard, USA.
25. Thomas Carlyle coined the term dismal science to describe economics. He is said to have been inspired by T.R. Malthus' gloomy prediction that population would always grow faster than food, dooming mankind to unending poverty and hardship. See https://www.investopedia.com/terms/d/dismalscience.asp

26. Liberto, D. 2022. Who is Joseph Schumpeter. *Investopedia.* https://www.investopedia.com/terms/j/joseph-schumpeter.asp. Quote is from P 71 of Schneider, H. 2017. Creative Destruction and the Sharing Economy. Edward Elgar Publishing, UK.
27. Harford, T. 2023. What Mystic Meg can teach us about economic forecasting. Financial Times 24th March. https://www.ft.com/content/cb9a6bfc-ff59-4879-9ac8-602c205d4f9b?xnpe_tifc=OF_XbubZx.xpbdnjbIYdhMpJVdUZMds_OuYlhunXbdn8tfxX4DxA4kesxCJNbDH8tfU_4d4uhIL_hfnphMpN4._J4Ixp4FzdxFYJOFLu&utm_source=exponea&utm_campaign=B2C%20%7C%20Subs%20%7C%20Weekend%20Campaign%20%7C%20Manual%20%7C%20250323%20%7C%20Test%20B&utm_medium=email
28. Petroski, H. 1990. The Pencil. Alfred A Knopf, New York.
29. Christopher Bullock is quoted as saying this in 1716. See https://blackalliance.org/death-taxes-quote/
30. Johansson, A., C. Heady, J.M. Arnold, B. Brys and L. Vartia. 2008. Taxation and Economic Growth. OECD Economics Department Working Papers No. 620. Section 1.1. Main findings states that during the past three decades there has been a reduction in the share of tax revenues accounted for by personal income tax while the revenue shares of corporate income taxes and social security contributions have increased. The share of consumption taxes in total revenues has declined, with the mix of taxes on goods and services changing noticeably towards greater use of general consumption taxes (mainly VAT) and away from taxes on specific goods and services.
31. OECD Economics Department Working Papers No. 620 (2008). Taxation and Economic Growth. Paragraph 49 page 21 states that owner-occupied housing has a favourable tax treatment relative to other forms of investment in many OECD countries through reduced tax rates or exemption for imputed rental income, mortgage interest payment deductibility and exemptions from capital gains tax. While the favourable treatment of owner occupation is often justified by the specific nature of housing... they may distort the flow of capital out of other sectors and into housing. They can also reduce labour mobility and thus the efficient allocation of labour. In these circumstances, raising taxes on immovable property could improve economic efficiency and growth. See https://dx.doi.org/10.1787/241216205486

Chapter 3

Rational People are Irrational

Nothing is Perfect—Excellence is Purposeful Progress through Learning

War does not determine who is right—only who is left.
You do not need a parachute to skydive. You only need it to skydive twice.

Two Sandwiches

Two sandwiches are laid on a table. Each sandwich has a paper label on it. One label read 'poisoned' and the other 'safe'. A blind man is desperately hungry and wants to eat one of the sandwiches. He is told about the labels but not which one is which. He is so hungry that he quickly chooses to eat one. If he chooses the poisoned sandwich, did he choose to die? Is he responsible for his own death? Would that also be so if he hadn't been told that one sandwich was poisoned?

Window taxes

At the end of the 17th century, there was a tax on the number of windows in a house. The idea was to make the wealthier pay more (in those days an income tax of the sort we have today was difficult to impose). Windows were a luxury so there was a high correlation between wealth and the number of windows a person had. The unintended consequence was that many householders bricked up their windows. Taxation changed behaviors.

A Man walks into a bar *and asks for 6 whiskies. He lines them up and proceeds to drink the 1st, 3rd, and 5th glasses. As he starts to walk out the barman says, "You haven't finished your other drinks". The Man replies, "I have a heart condition and my doctor told me that I could only have the odd drink".*

NOTHING IS PERFECT

Humor goes beyond rationality—it counters our presumptions, how things should be. It can help to diffuse tense situations and connect people. I recall a cartoon that was hanging on the laboratory wall of my colleague Dr Bill Larnach back in the 1980s. Bill was a specialist in foundation engineering—the engineering of the ground to support buildings and bridges. He had a strong sense of humor. The cartoon showed a vertically straight Tower of Pisa in Italy—now of course famous for its lean. Two men in Italian medieval clothes were talking. The captions read, "*I skimped a little on the foundations—but no-one will ever know*".

Perhaps the medieval engineer thought he was behaving rationally because he considered that the foundations were too big. Perhaps he could make money by reducing quantities. Of course, we now know that he was wrong. In 1990 a team of modern engineers began work to stabilize the tower and prevent it from falling over. They succeeded, now tourists can walk up it again. But was our medieval really engineer wrong? If the Tower had remained straight, it wouldn't have become a money-spinning tourist attraction. Rationalism seems to depend on perspective.

The engineering story was also an interesting story of engineering ingenuity. Professor John Burland, a friend and colleague, working at Imperial College, London was asked to be part of the 14 member Italian engineering team. John told me that he got a phone call from an Italian engineer friend. When he asked, "*How are you*?" his friend replied, "*I was fine until I opened today's newspapers and found that I was chairing a committee to decide how to prop up the Tower of Pisa*". "*Commiserations*", said John. "*Save them for yourself*", came the reply, "*the paper names you too*!" The formal invitation came from Rome much later. The full story is told on the tower's own website[1]. It is impossible to imagine the weight of responsibility in being asked to intervene in one of the world's most famous tourist attractions. Had they got it wrong, and the tower had fallen, you can just imagine the consequences. Stabilizing the tower was a unique very complex and highly uncertain challenge with no guarantee of success.

John worked for 11 years on the project. The team eventually decided to remove some 38 cubic metres of soil very gradually through long tubes inserted beneath the raised end of the tower. The Tower slightly straightened to the way it had been in 1838. Unfortunately, the Tower continued to lean a little more each year. They found that the water table was higher on one side and in the winter rains the water table rose. The problem was finally solved by installing drains and in May 2008 the engineers announced that the tower had been stabilized. In April 2011, the scaffolding came down to reveal clean restored stonework. John was awarded the Italian Knight Commander of the Royal Order of Francis I by the Duke of Castro for his work.

John Burland's engineering was totally thought through using his deep understanding of possible ground movements under the tower. Nevertheless, he experienced emotional and difficult moments as some of the critical results of the measurements of the movement of the ground emerged.

Rationality works—but sometimes we have to be rationally ignorant. This is because you can never know everything and therefore sometimes you have to choose to ignore certain things. It saves unnecessary time and effort. For example, when you buy a new TV, you don't always need to find out everything you could discover about the type, brand, materials, and methods of manufacture. Just some basic information is enough to decide which one to buy. However, if you have the same attitude to your food, then you may become obese if you choose what to eat badly and in ignorance. Health and diet are importantly linked. Being rational to you may or may not be seen to be as rational by someone else. For example, arranged marriages were once the norm in 18th and 19th century Britain, and still are in some countries such as India and Pakistan. Land, property, and social status were the important criteria not individual choice and love. What was once rational in the UK is now unthinkably irrational because our criteria for a good marriage have changed.

In everyday life we feel that we can choose our actions and so we may feel we are choosing rationally. This is not how it is in practice. The stories at the head of this chapter may sound a bit far-fetched, but they are not far from what happens every day when we make decisions that seem rational but in practice are harmful to ourselves and others. To others those decisions may seem irrational.

Often, when we make irrational decisions, we didn't intend any harm, such as when we tell a 'white lie' to spare someone feelings. But we may be blamed when the lie is exposed. Others assume that we know exactly what we are doing. Some are quick to blame and justify themselves by saying we did something on purpose. They feel justice must be done and misdeeds punished. This is inaccurate. The multiple ways in which we make decisions, often change our minds, and act upon those decisions can be both rational and irrational depending on who is affected. If we are thinking logically and considering possible consequences, then generally we are being rational. If we rely on our emotional reactions to events and possibly ignore our past experiences and common sense, then we may be being irrational. Certainly, John Burland's thinking was rational based on his long experience in ground engineering—but that didn't mean that his reactions to developments were unemotional. One thing is for sure—we can't afford to choose to be rationally ignorant about important concerns such as fake news and climate change.

OUR RATIONAL ECONOMIC DECISIONS ARE NOT ALWAYS RATIONAL

My learning points 1 and 5 in Chapter 1 were that one of the three surreptitious assumptions at the heart of capitalist economics is the idea of the *'rational'* economic man and the notion of usefulness or utility. An economic man (*homo œconomicus*) is an idealized person who acts rationally, with perfect knowledge, to maximize personal satisfaction[2]. It is an assumption at the heart of many economic models, and I maintain set economics off in a seriously wrong direction.

We need to examine these learning points a little more deeply. As we know rationality has brought us many good things—but it is not the whole story, far from it as any practitioner will tell you, and there is a dark side too.

On the good side, common sense tells us that whilst self-interest is important, people often act in ways that are not always in their own rational self-interest. Acts of bravery, acts of caring and acts of sheer kindness are common examples. People are not just only concerned, as many economists assume, to maximize benefits and minimize costs. We know that in everyday life we have often to make decisions with insufficient knowledge, feedback, and ability to understand and process the information that we can access. Some call this 'bounded rationality'—but whatever we call it, it is obviously highly uncertain. On top of that, sometimes we lack sufficient self-control, our emotions rule us, for example, when we buy something on impulse, and our preferences change—often because context changes the situation has changed.

At root we are all trying to make sense of our world so we can act in a way that is rational to us. To do that we feel a need to understand to give meaning to our collective experiences. Through that process we create some kind of purpose for ourselves. For example, we love being creative.

There seem to be six interacting aspects of the ways in which we create meaning. The first is the way we think about ourselves—our identity, sense of self, our beliefs and our very being. Second is reflection or what we notice and give attention to—what interests and motivates us. Then comes narrative or the stories we make to help us understand and simplify complex issues. Fourth we share our stories with others—the social aspects are foundational. Fifth, sense making is dynamic and ongoing and will change. Finally, we extract clues to decide what is relevant and acceptable. We tend to look for plausible explanations that 'fit' our worldview rather than spending time searching for ones that are accurate.

Many of us do not see ourselves as creative because we are told that creativity is a special gift given to a few—the artists who draw, paint, and make music. Of course, artists have special talents. But we can all be creative when we make sense of a situation and formulate a problem. When we formulate our problems we define needs, operators, constraints, and criteria. That constitutes a system—and it is not necessarily obtained logically.

Daniel Kahneman[3] uses a metaphor to capture the way we do some things on autopilot whilst others take time. He says we have two systems. System 1 is fast, automatic, emotional, stereotypical, and unconscious. For example, riding a bicycle on an open road or displaying disgust at a macabre image. System 1 associates the new information with existing brain patterns. It doesn't create new patterns and it explains why we are often biased in our judgements. Kahneman's System 2 is much slower and requires effort from us. It is logical, calculating, and conscious and we use it less frequently than System 1. For example, doing some arithmetic or focussing on something that needs your attention. When we use System 2, we form new brain patterns, and that is when we learn something different. Kahneman's analysis suggests that we don't make rational decisions based on outcomes. Rather we look at gains and losses based on heuristics. That means we often make sub-optimal choices. We are irrational agents by nature. Our brains have cognitive biases that commonly stop us from seeing the world as it really is and interacting with it in a way that will best benefit us. Many of these biases derive from our emotions. We may be too quick to trust our intuition.

This has led many people to see the way we sometimes jump to conclusions, often out of context, with goals that conflict our considered logical thoughts, as a war. And it's not a war we always win.

Lisa Feldman Barrett[4] takes different but similar view. She sees emotions as complex constructions, not simple circuits in the brain. She rejects the idea that our brains are a battleground between reason and emotion—thinking and feeling. She says that emotions are not built-in patterns of neurons in our brains rather they are built by us. For example, if we see an image, we immediately try to match it to something in our experience. We attempt to categorise it into a concept that makes sense to us. If we can't do that—she calls it 'experiential blindness'—we can't identify the image and so cannot give it any meaning—it makes no sense to us. So, it is experience that forms the knowledge that helps us identify what we see. Our brain is making sense of information by categorising it. It is making sense of sensory inputs from the world and your body to construct experience into a category to construct a concept as an emotion. Physical changes in your body have no inherent emotional meaning. Your brain makes bodily changes meaningful as an emotion or feeling. So, she is saying that emotions don't simply happen to you. They are made by you. Emotions you read in other people are coming, partly, from inside your own head. Your brain is guessing what that physical signal means in context that includes your surroundings but also your own body. She makes the crucial point that you are the designer of your own experience and hence of your own actions. You cannot change that categorisation in an instant, but you can influence it over time. Barrett's theory is one of constructed emotions in an emotional landscape. And it is based on survival. The landscape gives us quick, succinct information about our circumstances so we can work out what to do next. The landscape is fluid and flexible as long as we don't confine it to our own narrow experience and culture. If we can educate our emotional landscape, then we can change. We can move our minds to align with any model we want for ourselves. That is what is called emotional intelligence—the ability to perceive, control, and evaluate emotions. And it can be learned—it's not inborn.

SCIENCE IS A FORM OF SENSE MAKING

What is the role of science in all of this? At root science is simply a way for making sense of our world. But one that is formalised. The so-called scientific method is a process of creating and testing ideas which we can then share, criticise, improve, and consequently rely on to some degree when we decide what to do. Science is not some infallible truth. It is ever changing. But science represents the best of our shared current knowledge and understanding at any point in time. It contains much uncertainty and is incomplete—we can't know everything about everything. So, we must treat it as tentative. But large parts of it are highly tested and therefore highly dependable as long as we use it in the context in which it was developed. For example, we can depend on Newtonian mechanics to build big bridges but not to make quantum computers.

A scientific theory is usually defined as a general observation that connects properties of observables or concepts built on observables. It is a model—a set of hypotheses or propositions that explain a group of phenomena. In classical mechanics an observable is something that can be measured. More generally it is something that can be categorised. It is of the form everything which is A is B. If A is a moving snooker or pool ball that strikes another ball B, then B will move. If the theory is thought to be true it is called a law. Further hypotheses can be deduced based on those already made. The hypotheses are layered from the topmost general such as the three laws of mechanics formulated by Sir Isaac Newton, to laws that follow logically (can be deduced) from them such as the theory of elasticity of materials. These are the laws and hypotheses are models that enable us to predict how the snooker balls may move.

There are three forms of rational logical thinking—deduction, induction, and abduction. We are using deduction when we can assert everything which is A is B and A is true then it follows that B is true. For example, A might be if the sun is shining the air is warm and B might be that the sun is shining. From those two statements A and B we can deduce that the air is indeed warm. But that begs the question of where does the first hypothesis about A and B come from? Through most of history scientists and philosophers thought it came via induction. Induction is a form of reasoning that takes us from many specific cases to the universal—it is the method of basing general statements on accumulated observations of instances. For example, the sun sets every day therefore the sun will always set. The philosopher David Hume (1711–1776) pointed out that no matter how many times we observe that an event *A* is followed by an event *B* it does not logically follow that *B* will always follow *A*.

A third method is called abduction. This is making probable conclusions from what we observe. Abduction allows us to assume that tomorrow the sun will set as usual. So, we rely on it for common sense reasoning all of the time. Basically, abduction involves forming a conclusion from the information that is known. For example, it is used as explanatory reasoning in justifying hypotheses. Some philosophers refer to it as 'Inference to the Best Explanation'.

Two of the most influential philosophers of science of the 20th century were Sir Karl Popper and Thomas Kuhn. Popper argued that science is an evolutionary problem-solving process—and the principal problem is survival. He argued that truth is not manifest and not easy to come by. To search for it we need imagination, trial and error and critical discussion. We need to be both rational and irrational. But we can never know the absolute truth about the world—all we can do is test our theories, act on them, and change them. We can know when they are false because if we deduce a statement from them, we can test and find to be false it follows deductively our theory is wrong.

The sciences, such as mechanics, thermodynamics and electromagnetism have succeeded in bringing us long span bridges, tall skyscrapers, combustion engines and space travel as well as mobile phones and computers. The reason is that all of the theories we have evolved can be tested very precisely in the rigorously controlled environments of a laboratory.

The reason why the social sciences are so difficult is that controlled testing is either difficult or impossible. Human intentionality is inherent—it plays a crucial role. Intentionality is our consciousness that gives us the capacity to be aware of 'states of affairs' outside of ourselves. It gives us the capacity to act purposefully (with intent) and commit to act deliberately. People have individual will and purpose which are complex and ever changing. If a cannon ball had intentionality, it could, part way through its flight, decide to change its trajectory. The fact that it doesn't allows us to predict where it will land. Modern missiles can be targeted quite accurately[5] because they have control systems that will use radar or heat signatures to find its target.

RATIONALITY DOES NOT PRECLUDE IRRATIONALITY

The idea that human beings are economic rational creatures stems from the Ethiopian theologian Zera Yacob[6] and the European enlightenment of the 18th and 19th centuries and thinkers such as Rene Descartes, Gottfried Leibnitz, Jeremy Bentham, and John Stuart Mill. They wanted to bring natural science to bear on all types of knowledge. Later we shall see that the modern school of thought called 'behavioral economics', challenges this notion. These early thinkers recognised that humans do have emotions and motivations outside of pursuing personal satisfaction, but they thought that they should be ignored. They wanted to strip a human being to the essentials to get to a central truth. Unfortunately, that leads to an economic man who does not have to act morally or responsibly. He doesn't even have to act rationally as seen by an outside observer—just to act to get his own goals at lowest cost.

Gerschlager[7] has analysed Adam Smith's moral and economic theories. She writes that '*an economic man is characterized by self-interested goals and a rational choice of means*'. Smith maintained that rational economic behavior is action based on the rules of logic and so violations of self-interest are failures of rationality. He is often interpreted as implying that an unconstrained economic man can cheat and deceive to gain advantage but as pointed out earlier, we know that is a misconception. Common sense tells us that if we know that someone may cheat on us, we are not going to trade with him or her. So, it is in a person's interest not to cheat. But this assumes that all parties to a trade are fully informed. Unfortunately, when information is asymmetric (unequal between parties) then cheating can be profitable at least in the short term. Cheating occurs when a person misrepresents information to persuade another to act in a way that he/she would not have chosen had he/she been fully informed. We all want to feel good about ourselves so most of us tend not to want to cheat. But as Dan Ariely[8] has pointed out many of feel we can cheat a little bit and still feel OK. I suspect that very few people have never cheated on their expenses, for example, by claiming a full car mileage allowance. However, there is a threshold level—that varies between people—that we won't go over. There are people who would never cheat if they can avoid doing so. The levels are intuitive and differ, so we often disagree what is a best course of action. Ariely says we need to find ways to test our intuitions rather than simply assuming we are right and the other person wrong. Regulations, contracts, and trading rules are examples of ways we try to remove cheating. They are designed, as far as possible, to make it irrational for a party to cheat. In other words, cheating is designed out as far as possible. That is why, when contracts are not well written, one of the parties to a contract may cheat. Indeed, unscrupulous people make a living by finding loopholes in regulations and contracts to get an advantage. They argue they are doing nothing illegal. But they may recognize that what they are doing is immoral because it goes against their obligation to respect the principles of right conduct.

Gerschlager [7] concludes that Smith's comprehension of social life, and thus of deception, advocates a broader view of economics, one that goes beyond the image of rational man. Deception is inevitable in society, and Smith advocated, she says, a view that is much wider than is generally recognised by modern economists. Smith's theory of self-love is based on a notion of sympathy as a '*power of the imagination which allows a person to put him/herself in another's place*'. One human being cannot know another directly—only indirectly. So, exchanges are an exercise of imagination—but self-love can cause self-delusions. For example, a vain man needs others to praise him. His vanity is reinforced by social interactions, and he is unable to judge impartially. Individuals are motivated not only by self-love but also by self-delusion—both embedded in social interdependencies. As we interact with others as we create our own identity. Smith thought that deceptions are individually but socially constructed and inevitably part of any exchanges. If we ignore Smith's wider social views, there is no point in constraining it. That idea found an ideal breeding ground in the marketplace. Some people saw it as justifying self-interest and greed. It gave them an excuse to ignores our obligations to the interests of others—including our environment.

One of the exemplars of modern rationalist thinking was Greg Becker—a leader of the third generation of the Chicago school of economics mentioned in Chapter 2. Many eminent economists worked at the Chicago school including Jacob Viner, Theodore Schultz followed by Milton Friedman and George Stigler. The Chicago movement was central to the development of the neoclassical synthesis of free markets with minimal government intervention. Those theories included monetarist policies to keep the money supply in equilibrium with the demand for money. Becker looked at sociological issues through his 'economic spectacles'. These included racial discrimination, crime, family organisation, and rational addiction. Unsurprisingly he decided that many different types of human behavior can be modelled using economic modelling—he saw people acting rationally to maximize their utility. He even claimed to be able to include those actions that are often regarded as self-destructive or irrational. Rather vaingloriously his interpretation of irrational behavior was very limited—he defined it as any deviation from the maximisation of utility[9]. He considered altruism which he said sometimes have self-serving ends if the utility of an individual is properly defined and measured. Nevertheless, he did understand the importance of human capital. Becker thought that public policy should be based on preferential choices determined by incentives.

Becker put all his faith in reason and human logic despite recognising irrational behavior is real. He seemed to overlook that, even if that process is correct, it is still only a model of reality and not the actuality. The universe is a complex system so that we can never be sure that we can incorporate every single detail of this system so that it is infallible. It is inevitably a limitation of our thinking minds. Rationality is one of the most valuable life tools, but alone, it's not enough. The best way to make decisions and to act upon them is neither rational nor irrational—it is a combination of both—a synergy.

Barry Shwartz[10] sees rules and incentives as blunt instruments for affecting behavior. He wants a focus on something that cannot be mathematised or put into an algorithm—character. He builds on Aristotle's notion of practical wisdom[11]. Practical wisdom is the will to do the right thing with the skill to work out what the right thing is in a context. He sees it as a kind of moral jazz where participants follow a schema but can extemporize to fit the situation where strict following of rules would obviously not be common sense. He argues that in modern life where rules and regulations abound, and where people tend to sue you if you put a foot wrong, we are required to follow procedures that common sense tells us are wrong. He gives many examples such as a child being taken from its parents for a minor breach of regulation with consequences that are more harmful than the original situation. A jobsworth is someone who says '*I can't do that, it's more than my job's worth*' implying that it is against the rules set down for him and would cost his job if he disobeys them. He seems to be deliberately uncooperatively to uphold petty rules. Other examples are health and safety rules that prevent children from playing with conkers unless they are wearing safety goggles or hanging flower baskets being banned in case people bump their heads on them[12].

Shwartz lays emphasis on choosing between doing what is the right thing and doing the expected thing or required thing or the profitable thing. When things go wrong, we feel that we cannot trust the people we depend on because they don't know us well enough. There tend to be two responses to this. One is let's make more rules, the other is let's make cleverer and better incentives. For example, rather than let judges use their judgement we give them rules for sentencing. When bankers misbehave, we regulate them more. We require teachers to teach to a set of rules rather than allow them to decide what is best for their pupils. But there is no set of rules, no matter how detailed, where clever people cannot find new ways around them if they decide to. Shwartz says what we need is not more rules and incentives but virtue and character. We need people who want to do the right thing. We need them to have practical wisdom—the moral will to do the right thing and the skill to figure out what the right thing is. We must allow them to bend the rules and improvise novel solutions to novel problems. We need to give them the flexibility to know when and how to bend the rules in the service of others. As I said earlier it's rather like improvisation in jazz music where musicians dance around the rules of the fixed notes.

Rules and incentives do not tell you how to be a good parent or spouse or a good doctor, engineer, or economist. They demoralise the professionals and their activities and create people who only behave to the incentives. But some people are what Shwartz calls 'canny outlaws' who find a way around the rules. For example, a teacher who follows the rules but finds time to do what he considers the right thing. But it is systems thinkers who seek to transform the system. To help them develop we need to nurture practical wisdom through exemplars and role models so that they are allowed to do the right thing by others and by ourselves. The key to happiness, says Shwartz, is love and work. That means good close relationships with others and meaningful occupation.

WE HAVE NO CHOICE BUT TO EMBRACE UNCERTAINTY–PURPOSEFUL PROGRESS THROUGH LEARNING

Nothing in this world of ours is perfect. Embracing uncertainty isn't easy but if we see excellence as purposeful progress through learning we can cope. But how do we classify uncertainty? Synonyms include truth, sureness, freedom from doubt or reservation, trustworthy, reliable, or dependable, unquestionable, inevitable, indisputable, definite, particular, or unambiguous. We can unpack these meanings by noting that there are three groups of kinds which I will later discuss in some detail. First is the age old philosophical and practical question of what we mean when we say something is true (truth, sureness) or not. In this I will include how we give 'meaning' or how we interpret what is expressed or indicated by a statement which may or may not be true. Second is the need to be able to trust what we say (freedom from doubt or reservation, trustworthy, reliable). Third is the level of clarity of what we say (definite or unambiguous).

Dictionary definitions of the word uncertainty tend to introduce other (opposite) terms like variable, erratic, unpredictable, indeterminate, sceptical, hesitancy, not assured or fully confident and ambivalent. Uncertainty is also described as a quality of being with respect to duration, continuance or occurrence, a liability to chance or accident. There are eight headings for the word uncertainty in Roget's International Thesaurus. They are irregularity, changeableness, indistinctness, unsureness, equivocalness, irresolution, fickleness, and precariousness. We can distil these many further aspects into three more kinds of meaning which are changeableness (irregularity, variable, erratic), incompleteness (indeterminate) and risk (precariousness, unpredictability).

These six kinds of uncertainty (truth, trust, clarity, changeableness, incompleteness, and risk) are manifest when we make decisions. Truth and changeableness (together as one kind) with clarity and completeness are three central attributes of information and trust is paramount in the making of rational decisions. They are all ingredients of managing and controlling risk.

COPING WITH UNCERTAINTY REQUIRES US TO BE AWARE OF OUR OBLIGATIONS

In Chapter 1, I said that the term economic man, as widely used in economic models, sets it in a seriously wrong direction. In our increasingly uncertain world, I called for a paradigm shift from the selfish rational man to the man of obligation—a change from the consumer story to the citizen story. Now it is time to explore the notion of an obligation in a little more detail.

An obligation is a binding promise, contract or sense of duty that ought to be used instead of utility. If we accept that rules, regulations, and incentives only take us so far then the rest is down to character. We need to develop practical wisdom through a sense of the obligations we have to each other and to planet Earth. Good faith and goodwill are to act on accordance with our values of legitimate standards of trust, honesty and upheld by the law. The philosopher Tom Settle[13] argues that obligations are an underived social relation—that is an obligation cannot be analysed in terms of some other concepts more basic than it, and from which it can be derived. He sees an obligation as a root reciprocal moral principle for any society of more than two autonomous rational beings. Values are the things we hold as important in the way we live. In short, the worth of what is the 'good'. Values determine our personal priorities such as education, wellbeing, honesty, truthfulness, loyalty, open-mindedness, consistency, and our corporate values such as sustainability, efficiency, creativity, ingenuity, and innovation. To act with good faith is to act reasonably. This implies that consent is not withheld by any of the parties unreasonably, depending on context. Under the law of tort, a duty of care is owed by each one of us towards any other to adhere to a reasonable standard of behavior that will do no harm. It is an expression of the social contract by which we live and work together.

Tom Settle sees an obligation as a root reciprocal moral principle. It is a principle found in the 'golden rule' of all religions and expressed in the Bible as—*'do unto others as you would have them do unto you'* (Luke, Chapter 6, v 31). Settle quotes John Dewey as pointing out that no genuine moral principle prescribes a specific course of action. But they do require a particular attitude and respect, such that we can try to imagine what it is like to 'stand in the shoes' of another. They also make clear what is not to be done.

As we have learned, especially in the last 20 years or so, the importance of our natural environment, we have begun to realise we also owe a duty of care to the planet and the natural world. Natural assets for an entity are part of the commons and are of three kinds: (a) used but taken for granted and not accounted—such as air and water quality. Damage reparation should be set against income but often is not; (b) used, recognised, and accounted as delivering ecosystems services; and (c) used and positive for the commons but perhaps detrimental to the entity. For example, eagles on a grouse shooting estate are an asset to the commons but a cost to the estate if it kills grouse.

I have postulated[14] that an obligation is the 'stuff' in a field of human interactions. It is analogous to the 'stuff' of gravitational and electromagnetic fields of science. A field is simply a region of space that is under the influence of some agency. In classical science the influence is one of experiencing a force from gravity or electromagnetism. The 'stuff' is a mass or an electrical charge in motion. In the usual financial system, the 'stuff' is the financial exchange worth (FEW) of part of the assets of one financial entity (the buyer/investor) for something desired to be supplied by another entity (the supplier). The objective of each entity is to increase its net worth by generating a net profit of income over expenditure through time. FEW is the relative state of being useful or important enough to be desired, valued in a manner that is trustworthy and considered dependable and reliable by all entities in the system. For the most part FEW is a form of money as currency (cash and non-cash), such as the £ or $, but in theory, it could be any form of trustworthy barter.

I suggested that rather than think of FEW as the stuff of trading we should use a different concept that I called 'The Worth of Obligation' (TWO). The Worth of Obligation entails Financial Exchange Worth (FEW). In other words, TWO goes beyond FEW. It recognises that whilst usefulness or utility is important our obligations to each other and to our natural world are more fundamental. I suggested that although TWO should be exchangeable it should not be tradeable in the same way as FEW and there should be no market for TWO that is not FEW[14]. Some things, such as the water we drink and the air we breathe—are too important to be valued as FEW. Of course, there are ecosystems services to which we apply FEW such as utility companies that collect, store, clean and distribute water to our taps. The water collected by the utility companies is free but delivering potable water to our taps is not. Likewise, the worldwide scandals of poverty, inequality and social deprivation are too profound and all-encompassing to be valued as FEW. Indeed, it can be argued that they derive from an improper use of FEW that we need to correct. The natural capital we take for granted but which is essential for life on earth must be measured by some other metric. We need a metric that does not depend on trade-offs susceptible to the tragedy of the commons or the prisoner's dilemma[15]. A new unit of TWO has to capture the degree or amount of worth of the moral obligations we owe to each other and to the natural world.

I have proposed that the new metric should be a new unit that I call *obligs*. An *oblig* is a way of capturing the amount of the substance of the 'stuff' that moves around in a social field of obligation forces. In other words, as we bargain and trade with each other and our environment, we make commitments to exchange that which we consider, at the time, to be equivalents. When we trade by only exchanging FEW, then TWO *obligs* become de facto units of currency as $ or £ and their reciprocal legitimate benefits as goods and services, as a special case.

Financial entity								
Assets			**Liabilities**			**Net Worth**		
Current	*R*	£	Current	*D*	£	Current	*R* - *D*	£
Non-Current	*B*	£	Non-Current	*U*	£	Non-Current	*B* - *U*	£
Biodiversity	*b*	*obligs*	Biodiversity	*b*	*obligs*	Biodiversity	*0*	*obligs*
CHG	*GHC*	*obligs*	CHG	*GHC*	*obligs*	CHG	*0*	*obligs*
			BioDamage	*d1*	*obligs*	BioDamage	- *d1*	*obligs*
			CHG Damage	*d2*	*obligs*	CHG Damage	- *d2*	*obligs*
Assets = *£(R + B) + obligs(b + CHG)*			Liabilities = *£(D + U) + obligs(b−d1 + CHG − d2)*			Net Worth = Assets - Liabilities		

Figure 3.1 A Balance sheet showing both FEW £(UK) and TWO *obligs*.

Figure 3.1 shows a balance sheet in which FEW and TWO are accounted for separately. In effect, TWO connects value (as financial worth) with values (as that which we consider important in the way we live) as Carney[16] advocates. TWO takes us away from inward-looking selfish individual usefulness to outward-looking collaborative obligation that entails a much wider set of values than finance. TWO is a 'bigger idea' than FEW with an importantly different moral stance from that of economic utility. Whereas utility is the usefulness and degree of satisfaction of a consumer's choice, TWO is centred on a set of obligations that a financial entity has to the natural world and to those aspects of our human and social world that are beyond FEW. The idea that utility can be the basis of moral choice as proposed by Hirsh, Lu and Galinsky[17] is flawed since it assumes that ethical choice depends on usefulness rather than a duty of care for the natural world and others. However, TWO obligations are not necessarily reciprocal. Although we human beings have obligations to nature—it has no obligation to us. Nature will do what it does regardless. It takes no account of human sensibilities, but we need to treat it carefully because we are part of it. We are taking for granted the air we breathe and the water we drink at our peril.

Humor is just one example of our joy in irrationality. Most of us feel that because we can choose our decisions, actions, and behaviors then we are choosing rationally. We recognize that there are aspects of life where our decisions may not be entirely rational because they involve our emotions, gut reactions, and feelings. What may seem perfectly rational to us individually may be harmful to others, ourselves and to the natural world. To others our individual rationality may seem irrational.

Common sense tells us that whilst self-interest is important, people often act in ways that are not always in their own self-interest. Acts of bravery, acts of caring and acts of sheer kindness are everyday examples. People are not just only concerned to maximize benefits and minimize costs as economists seem to think. We know that in everyday life we have often to make decisions with insufficient knowledge, feedback, and ability to understand and process the information that we can access. Some call this bounded rationality. Often, we lack sufficient self-control, our emotions rule us, for example, when we buy something on impulse.

Daniel Kahneman[3] suggests that we don't make rational decisions based on outcome, but rather in terms of gains and losses using mental heuristics that often lead to sub-optimal choices. We are irrational agents by nature. Our brains have cognitive biases that stop us from seeing the world as it really is and interacting with it in a way that will best benefit us. Many of these biases derive from our emotions—consequently we may be too quick to trust our intuition. Lisa Feldman Barrett[4] sees emotions as complex constructions, not simple circuits in the brain. She rejects the idea that our brains are a battleground between reason and emotion—thinking and feeling.

Nothing in this world of ours is perfect. Embracing uncertainty isn't easy but if we see excellence as purposeful progress through learning, then we can cope. Science helps us decide. Science is a formal process of sense-making. It is the creating and testing of ideas that we can depend on to a degree in a context. From this kind of shared knowledge, we can design rules and incentives. But in practical reality, especially in finding trade-offs, they are blunt instruments. For example, they take no account of character—something that cannot yet be put into an algorithm. In treating our golden goose, we need to resurrect Aristotle's notion of practical wisdom—a notion lost in the mists of history. Practical wisdom is the will to do the right thing together with the skill to work out what the right thing is, in a context. It is a kind of moral jazz where participants follow a schema but can extemporize to fit the situation where strict following of rules would obviously not be common sense. It is an attribute particularly needed in a world where nothing is perfect.

Utility is central to economic theory, but it is selfish in the way it depends on preferences and incentives. In an uncertain world it should be replaced by the notion of an obligation to each other and the natural world. That obligation represents a binding promise, contract, or sense of duty. It is our duty of care towards the golden goose. The sense of obligation will help and enable us to deal with economic value as different from values as those things to which we attribute 'worth'. We must always hold this distinction in our minds. The economic value of a good or service, as a market value, is essentially determined by exchange or trade. It derives from its utility, usefulness, or benefit to a person and measured using money—units of currency. Essentially it is subjective and difficult or even impossible to measure in any absolute sense like the volume of a space[18] because it depends on the intentions of that person and their relationship with the good or service. Nevertheless, we have ways of estimating economic value. For example, producers set prices considering all sorts of tangible and intangible factors—including brand name. One way of estimating economic value is by examining that person's willingness to pay a price. That depends on preferences and incentives as Greg Becker stressed. It infers that their choices to purchase mean that they put a higher economic value on the good than on that amount of money. Economic value and market value can be quite different. Market value is the price set in a market for a good or service and can be higher or lower than the economic value that any one person puts on that good or service.

Values are the norms of behaviors and beliefs held by a person or societal group about what is 'good'. They embody how we differentiate between the bad and the good in a community, culture, or society. They include standards of behavior such as patience, honesty, loyalty, compassion, kindness, integrity, selflessness, and spirituality. They embody the common good and our obligations to each other. I have proposed a new unit of 'The Worth of Obligation' (TWO) which could be incorporated into a standard balance sheet to capture obligations to others.

Rationality has brought us many things from electric lights to mobile phones, from ploughing fields to combine harvesters, from bicycles to space rockets and from antibiotics to brain surgery. But our rationality can seem to hide our irrationality that brings us emotions and relationships with others and makes us essentially human. The gap between our rationality and our irrationality has brought us violence, crime, war, and conflict in many forms and now threatens our very existence as climate change begins to bite. The key to the role of economics in modern life is to know and differentiate economic value from the values we live by. So that is what we must turn to next—reversing our tendency to know the price but not the value of the things most important to us.

End Notes

1. *How was the Leaning Tower of Pisa stabilized?* 2022. See https://leaningtowerpisa.com/facts/how-pisa-leaning-tower-was-stabilized
2. Investopedia 2022. Economic Man. See https://www.investopedia.com/terms/e/economic-man.asp#:~:text=and%20accounting%20industries.-,What%20Is%20an%20Economic%20Man%3F,assumption%20of%20many%20economic%20models
3. Daniel, K. 2012. Thinking Fast Thinking Slow. Penguin, USA.
4. Barrett, L.F. 2017. How Emotions are Made: The Secret Life of the Brain. Houghton Mifflin Harcourt, UK.
5. Modern missiles have control systems that guide them to their target. Unlike a cannon ball they have a primitive form of intentionality. Cannonballs were solid chunks of metal that could be fired from a cannon with a range that can be estimated according to local conditions. Modern cruise missiles are like small pilotless aeroplanes with GPS (Global Positioning Systems), terrain contour mapping and inertial guidance with a camera to find its target.
6. Laudika yaNdangi. 2021. Zera Yacob, the Forgotten African Philosopher. See https://www.techgazi.com/zera-yacob-the-forgotten-african-philosopher/
7. Gerschlager, C., Beyond Economic Man: Adam Smith's concept of the agent and the role of deception, Cahiers d'économie Politique, 2005/2 (n° 49), pp. 31–49. 10.3917/cep.049.0031. https://www.cairn.info/revue-cahiers-d-economie-politique-1-2005-2-page-31.htm
8. Dan, A. 2009. Predictably Irrational. Harper, UK has pointed out many of feel we can cheat a little bit and still feel OK.
9. Becker, G.S. 1962. Irrational Behavior and Economic Theory. Journal of Political Economy. Feb. 1962. 70(1): 1–13. The University of Chicago Press. https://www.jstor.org/stable/1827018
10. Barry, S. 2011. Practical Wisdom. Riverhead Books, USA sees rules and incentives as blunt instruments.
11. Blockley, D.I. 2020. Creativity, Problem Solving and Aesthetics in Engineering. Springer. On page 29 I promote a principle of ingenuity which says '*value, nurture and develop practical wisdom*'.
12. The UK Health and Safety Executive (HSE) have identified 10 ridiculous myths associated with health and safety—See https://www.ihasco.co.uk/blog/entry/2162/10-ridiculous-health-safety-myths
13. Settle, T. 1976. In search of a third way. McClelland and Stewart, USA, and Settle, T. 1969. Scientists: priests of pseudo-certainty or prophets of enquiry? Science Forum. 2(3): 21–24.
14. Blockley, D. 2022. Exchanging obligations: accounting for all forms of capital. Journal of Interdisciplinary Economics. 0(0). https://doi.org/10.1177/02601079221081717
15. King, M. and J. Kay. 2020. The End of Alchemy. Abacus/The Bridge Street Press, UK.

16. Carney, M. 2021. Value(s). William Collins, UK.
17. Hirsh et al. argue that Moral Utility Theory provides an integrative framework for understanding the motivational basis of ethical decision making by modelling it as a process of maximising subjective expected utility. They see the 'warm glow' of altruism and taking pride in one's work as an intrinsic utility whilst guilt is an intrinsic cost of violating a moral norm and punishment is an extrinsic cost. Hirsh, J.B., Lu, J.G., Galinsky, A.D. 2018. Moral Utility Theory: Understanding the motivation to behave (un)ethically. Research in Organisational Behavior. 38: 43–59.
18. Measurements of the physical world have four particular attributes that make them dependable (Blockley, D.I. (1980). The Nature of Structural Design and Safety. Ellis Horwood, Chichester, UK). First, they are repeatable—you can measure the length of a line with a ruler many times. Second, the state of the phenomenon is unchanging—neither the line nor the ruler alters. Third, those states are clearly identifiable—you can see the divisions of mm and cm (or inches and feet) very clearly. Finally, the magnitude of change in the state of the phenomenon does not vary, i.e., it is repeatably the same—the line does not change its length between measurements. By contrast measures of phenomena involving people violate these conditions. This is because humans have intentionality. They may change their minds and their behaviors between measurements.

Part 2

Testing

Preamble to Part 2

I said in Part 1 that we become what we ***think***. Now in Part 2 we look we how to ***test*** what we read and are told. We want to discern and discriminate better the information 'wheat from the chaff'. We wish to distinguish information that we can rely on from that which is misleading or downright false. We need to help each other to test and separate fake news and on-line gossip from bone-fide facts and well researched opinion and reject rational ignorance that impairs us. It means revaluing how to discuss difficult issues critically with tolerance and goodwill and without taking umbrage.

As Mark Carney has observed so many of us have come to know the price of something but not its value. Often you don't know what you have got until its gone. Covid-19 was a reminder, if we needed one, that the everyday things we take for granted can be suddenly snatched away.

If we are to change and put money in its proper place so that economic thinking includes natural, human, social, cultural, ethical, and political capital as well as financial capital then collectively we must understand economics better. If we decide just to leave it to the specialists, how do we test what they are telling us? How do we know which politicians to believe when they pontificate on economic matters?

My learning point number 4 is that fiat money is not well understood by many people in power—especially politicians—and this leads to wrong-headed policy decisions. In brief fiat means by decree. Fiat money is a currency that is declared legal tender by a government that has no intrinsic or fixed value and is not backed by any tangible asset such as gold.

One of the most important pieces of information that few of us are aware of is the national balance sheet. UK national balance sheets are produced annually by the UK ONS (Office for National Statistics) and are available on-line. How many of us read and understand them? Assets are things that we own. They are financial and non-financial. Non-financial assets include land, buildings, equipment, and intellectual property which are not primarily financial but obviously do have financial value. In Chapter 4, we will examine how we measure things to put into the balance sheet and how reliable are the results as evidence. What is the nature of evidence and how do we cope with complexity? Statistical analysis is important in economics—it has its own sub-discipline called econometrics. It is that branch of economics concerned with the use of economic theory, mathematics, and statistical inference to quantify economic phenomena.

But in practice there is widespread use of indicators, often expressed as ratios, to understand, to learn and to get better at noticing things that are important and ignoring things that are less important. This is called 'selective attention' by some scientists. Through it we can (but often do not) learn from our successes and our mistakes. But this becomes difficult when reports of data and events rarely tell us about any assumptions behind the so-called facts.

It is often said that spending is the lifeblood of capitalism. We will examine the idea. Capitalism is an economic system in which private individuals or businesses own capital goods. The production of those goods and the services that accompany them is based on supply and demand in the general market—known as a market economy[1]. The main alternative system is the planned or command economy controlled through central planning as in Russia and China.

In the 2020s we have massive amounts of information thrown at us. In Chapter 6 we look at how we can sort out the 'wheat from the chaff' Fake news and sheer gossip via social media are doing immense harm. Examples include getting teenage children to share intimate details of themselves and even to commit suicide. We must not rely on fake news to vote if we want a healthy democracy. The internet has made it possible to share excellent information about specialist topics such as health, economics, and technology. It has transformed the publication of research. However, the complexity of systems interdependencies, and the lack of clarity on ethical issues are difficult to test adequately. Is there a 'right' thing to do in specific situation (for example, telling a 'white lie' to protect someone's feelings) and if so, can we test it out? Our present form of democratic or liberal capitalism has led some to greed and corruption. We need to change it to one of principled capitalism. We need to replace notion of utility and usefulness by moral rules that represent our common humanity whatever our divisions of religion, race, or gender.

The reasons for building a united common purpose for the flourishing of life on earth are impelling. Climate change will not discriminate. It is a common foe. Whether we live in a democracy or authoritarian state, whatever our race, gender, religion, or sexual orientation, whether we are wealthy or living in poverty, the level of increased storms and violent weather coupled with rising sea levels and loss of biodiversity will affect us all.

The military understand the importance of common purpose. Without it, armies are defeated. The soldiers define unity of effort as the '*product of successful unified action*' and consists of '*coordination and cooperation toward common objectives, even if the participants are not necessarily part of the same command or organisation*'[2]. Unity of effort is the glue that holds together the disparate actors as they pursue their collective goal.

Of course, a grand common purpose can itself be unjust if minority opinion is crushed or even treated in a patronizing fashion. Conflict then becomes inevitable. There is increasing acknowledgement the world over that what we need is not 'peace' alone, but rather 'peace with justice'. It is also why even war under certain circumstances has been deemed 'just', for example, in just-war theories.[3]

Nevertheless, climate change overrides everything else, but at the same time it presents us with a new imperative to find ways of understanding the human condition. It challenges us to find tolerance where there is intolerance, common interests where there is conflict, collaboration where there is disharmony, and communication where there is disjunction and deception. If we do not do these things, then we will be no more.

The warnings are serious. Our deep conflicts over natural resources, territory, religious and racial differences that have led to disagreements and bloody wars all over the world. They are tantamount to squabbling over the deck chairs on the Titanic as the climate change iceberg comes into view. The threat of that iceberg is existential.

THE NEED FOR GOVERNMENT INVESTMENT AND THE DEFICIT MYTH

Governments have been slow to invest because they worry about debt. As we will see through the lens of MMT that is simply wrong-headed economics. Stephanie Kelton has written her in her book,[4] 'The Deficit Myth' as her Myth No. 1 that the federal government (USA) should budget like a household. In reality, the federal government issues the currency it spends. Her myth No. 2 is that deficits are evidence of overspending. In reality for evidence of overspending look to inflation.

Economics is constantly in the news—a topic of regular discussion by the media and opinion formers and a major determinant of our choice of elected governments. But all is not well. The basic frameworks and principles of economics are no longer providing a way of adequately addressing the problems of today's rapidly changing world. Since it is we, the public, who vote in our politicians and who rely on our professional economists, we have a duty of care to understand what the specialists are saying and doing in our name. As I have already said, economics is too important to be left to economists and politicians. Whilst we must acknowledge their professional skills, knowledge, and expertise each one of us has a duty of care to each other and to the natural world, to have an informed understanding as we decide which way to vote.

End Notes

1. Wolf, M. 2023. The Crisis of Democratic Capitalism. Penguin Books, London. https://www.penguin.co.uk/books/305263/the-crisis-of-democratic-capitalism-by-wolf-martin/9780241303412
2. U.S. Department of Defense. 2023. Dictionary of Military Terms. Unity of effort is the glue. See also Kingsley, R. 2017. The Fundamental Principle of "Unity of Effort" in Multinational Operations http://militarycaveats.com/7-the-fundamental-principle-of-unity-of-effort-in-multinational-operations/
3. O'Donovan, O. 2012. The Just War Revisited. Cambridge University Press, UK.
4. Kelton, S. 2020. The Deficit Myth. John Murray, UK.

Chapter 4

Knowing the Price but not the Value

Some People are so Poor, All they have is Money

—Big Sean (American rapper)

The pigs were in charge at Animal Farm. *At a meeting of the Pig Council, the Chief Pig said, "We are getting overrun with rats—I think we should offer every animal on the farm a reward for every dead rat they bring to us". The Council agreed and a decree was sent out. The scheme worked well—many rats were killed and brought in. But the total population went up. There were more rats around the farm than before. When the Pig Council officers investigated, they found that the animals were breeding rats, then killing them just to get the reward.*

Beware unintended consequences and perverse incentives.

A Lion and a Man *were travel together through the forest. They began to quarrel. Each one boasted that he and his kind were far superior to the other both in strength and mind.*

Then they came to a clearing in the forest. There stood a statue of Hercules strangling the Nemean Lion.

"See", said the man, "that's how strong we are! The King of Beasts is like wax in our hands!"

"Ho!" laughed the Lion, "a Man made that statue. It would have been quite a different scene had a Lion made it!"

It all depends on the point of view, and who tells the story.

Man: *"You are 20 years old and engaged to marry a 90-year-old multi-billionaire. Are you marrying him for his money?"*

Woman: *"What a ridiculous suggestion—of course not"*

Man: *"Astonishing—I would never have guessed it"*

YOU DON'T KNOW WHAT YOU HAVE GOT UNTIL ITS GONE

Covid-19 was a reminder, if we needed one, that the everyday things we take for granted can be suddenly snatched away. Before the pandemic we could go to the shops whenever we wanted, we could eat out, play sports, have social occasions with family and friends, and go to meetings, performances, and games. We could sit close to other people and talk about all kinds of topics. During the forced isolation school children, who had previously hated lessons and being made to work when they would rather be out with their friends, were suddenly missing going to school. Parents who hated their work, but needed the money, were suddenly missing having to go to work. Retirees who enjoyed a strong social life were suddenly isolated. Worse still, parents and grandparents died without their seeing their loved ones one final time. Relatives had to grieve alone. We found ourselves regretting not having said the simple things like 'I love you' when we could have done.

As we get older, we are nostalgic about the good old days and wish they would return—perhaps forgetting the bad times and idealizing the good times. We remember old friends with whom we have lost touch and then we hear they have died, and we are full of regret that we didn't contact them again.

Perhaps rather than 'you don't know what you have got until its gone' we should say perhaps 'you don't know and appreciate the good things that you have' and don't give much thought to the fact you might lose them? Most of us tend to focus on the things we don't have rather than the things we do have. We take breathing clean air for granted—until we cough because its polluted.

Personally, as a retired University lecturer, I miss the feeling of looking forward to meeting in October of each year the fresh new students and welcoming back the older ones. In my first few years in the job, I was 'green'. I made mistakes. For example, I was once asked a question in a lecture, to which I didn't know the answer. Instead of saying I would find the answer for the next lecture I bluffed and got it wrong. At that next lecture I was extremely embarrassed at having to correct myself. Never again did I bluff—much better to suffer the indignity of admitting you don't know. It's a good lesson for students to find that even their professors are sometimes ignorant—they are human after all. Yet of course they knew that. I still miss that excitement of walking into a lecture room and the buzz of giving a half decent talk. Every term we used to get written feedback from our students as to how they felt about our teaching and lecturing. For many years I taught a class in engineering design for 3 hours every Wednesday. One student I remember, wrote on his feedback form *'I enjoy the class, but I have to say that 3 hours of Dr Blockley is too much for anyone—remember we students only have a short attention span'*. Over the years I have met many of my former students many years after graduation. It is a real pleasure to hear them speak so fondly of their time as students. It is a salutary lesson for me that in recent years I find that some of them are on the verge of retiring themselves.

Life's experiences are a constant reminder that we live in an uncertain, unpredictable, and capricious world. A flash of fate can suddenly transform the way we live. Severe crises like Covid-19 and the Ukrainian war threaten our very existence. They put big stresses on people, cause economic upheavals, hunger and shortages and extreme poverty. But they also bring out the good in people. Acts of benevolence, kindness, caring, help and empathy led to cooperation and a 'coming together'.

A common enemy makes us feel better connected to others. Covid-19 was a common enemy. But there is an even bigger one waiting for us—climate change. One that is no respecter of race, colour or creed. Whether rich or poor we are all affected. You often now hear people say, 'We must save the planet'. We must save the golden goose. But we should remind ourselves that no matter what the planet will survive albeit in a different form. What is at stake is all forms of life including we humans—the laying of the golden eggs.

In 1892 Oscar Wilde[1] wrote that a cynic was '*a man who knows the price of everything and the value of nothing'.* This nice turn of phrase hits at the heart of our problems. Economics and politics seem to be dominated by Wilde's cynics. The idea that real world activities should even consider ethical, moral, philosophical, or cultural values as on a par with economic value seems, to many people, as overly sentimental and romantic. We are implicitly, if not explicitly, told to be hard-nosed, rational, and practical. How many businessmen justify hard decisions that affect their employees or competitors by saying 'It's just business—no hard feelings'. Feelings are secondary to some economists and businesspeople. Yet love songs dominate popular culture and emotions prevail in much of the arts. In business finance teams are taught how to put together expense or tax systems into a spreadsheet to examine different scenarios of future outcomes. But many pertinent factors have no role. How do you include the effects of sexual harassment in a spreadsheet? The Fawcett Society[2], which campaigns for gender equality, says "*sexual harassment is...about a culture...which violates the dignity of...women...At least 40% of women have experienced workplace harassment*". In April 2023, the Confederation of British Industry sacked its director general after a series of sexual-misconduct allegations against senior figures. Harassment can't be ignored. And it wouldn't be if we placed an emphasis on the moral obligations of principled capitalism. In a similar vein, how do you include the need for and consequences of whistleblowing? What results from treating employees as commodities? Zero hours contracts, where officially workers are only paid for hours worked. Sometimes, unofficially, they are not paid annual leave or are not protected from discrimination. Moral obligations as a social norm in business would transform working conditions.

Hard-nosed rational financial teams may calculate a probability of getting a particular outcome, and then look at its downsides. But what are the unintended consequences they might miss? They perform calculations to find a best price to charge. For example, a publisher who needs to decide whether it's better to sell a small number (of books) at a higher price than a large number at a lower price. In publishing the former kinds of books are known as professional/technical and reference books and the latter as trade books. The same methodology might be used to decide personally whether to buy something cheap but with a greater chance of breaking, or something more expensive with a greater chance of lasting longer. Universities may decide to admit large numbers of students to maximize income from fees. But that brings the risk of lowering academic standards. They may decide on a small number of carefully chosen students who show a potential for high achievement—but with reduced income and all that entails. The methodology can even be used by some unscrupulous people to decide whether it's worth breaking the law. It will, in their minds at least depend on the probability of being detected and punished.

One of the common methodologies that gives little weight to moral obligation is called cost-benefit analysis. The benefits are simply compared with the costs of both actions and their various foreseeable consequences. The analysis requires measurable financial metrics such as revenue earned, or costs saved but it can allow, to some extent, for intangible benefits and costs such as the morale of employees and customer satisfaction. But only if they are transferrable into money.

More complex analyses may incorporate sensitivity analysis, discounting of cashflows, and what-if scenarios. Despite a nod towards intangibles the measuring and quantifying of things, underlying issues may be omitted. A focus on price makes it easier to miss value wider values—whether ethical, moral, cultural, empathetic. The whole set up is modelled as an exercise in numbers and games. Indeed, game theory is a branch of applied mathematics that enables a modelling of players making interdependent decisions.

One of the most famous examples is the prisoner's dilemma. Two criminals called Bill and Alf are arrested. The police have no hard evidence with which to convict them. So, they question each one separately making sure that they cannot communicate with each other. The police then have four scenarios to put to the prisoners. First if both criminals remain silent, they will each get a three-year prison sentence on a lesser charge. Second if one of them (say Bill) betrays the other (Alf) who remains silent, then Bill goes free, and Alf will get nine years. Third if the confessions are the other way round and Alf testifies against Bill who remains silent then Bill gets nine years and Alf goes free. If they testify against each other then both will get five years.

If you were Bill or Alf, which scenario would you choose? In overview the most favorable strategy, obviously, is not to confess, to remain silent and hope the other does likewise. But the dilemma is that neither Bill nor Alf know what the other will decide. Without that certainty (that the other will not confess), both may well testify against the other and hence both get a five-year sentence. Some experts argue that both players will choose what's best for them individually (going free) but worse for them collectively. But what if Bill and Alf were lifelong friends and one of the wives was expecting a baby soon—would he choose in the others best interest rather than his own. A tit-for-tat strategy suggests that it is better to assume that the other will cooperate so neither Bill nor Alf will confess.

Another more economic example of a cooperative strategy that works for the common good is for countries to decide not to impose import tariffs in the expectation that others will follow suit.

The key word here is game. If the only value is economic the problem becomes a game. A very narrow version of efficiency rules the day. Cutting costs and maximizing income to improve the bottom line becomes the key to almost everything in life—it has become a way of life, a mantra.

One example of how debilitating this can become was the scandal of the expense claims of UK members of parliament which was first exposed in 2009. Quite reasonably MPs can claim for the expenses they incur in doing their job. But there are rules governing what claims are, and what are not, acceptable. It is quite understandable that individual MPs would examine the rules to find ways to maximize the benefit they might get. But some MPs lost their sense of moral value and became greedy. The list of items that were claimed was eventually published by House of Commons resources chief. It included items such as dishwashers, television sets and tumble dryers. One MP employed his two sons, both full time undergraduates, as part-time salaried research assistants. Investigations found that the sons did no substantive work for their dad. He apologized, was suspended for 10 days, and eventually resigned. Another MP was found to have misapplied part of her parliamentary allowances and ordered to repay them. Other couples bought a second home and claimed mortgages costs. Taxes were avoided (legally) and attempts made to evade tax (illegally). And so, it went on. The disclosures of widespread misuse of expenses aroused much anger in the UK public and cause many resignations, sackings, de-selections, and retirements. Some MPs were prosecuted and given prison sentences. If we cannot count on our publicly elected MPs to fulfil their obligations to the people who elected them—then we have reached a very low point in our behavior towards each other.

In February 2020, the aircraft company Airbus agreed to pay a record 4$ billion in fines for alleged bribery and corruption over at least 15 years. The company reached a plea bargain (pleading guilty to a lesser charge for the dropping of other charges) with prosecutors in Britain, France, and the United States. According to prosecution documents, Airbus used a global network of agents or middlemen for corrupt transactions, included payouts disguised as commissions to push airplane sales. The bribery scandal reverberated around the world as investigations were launched in countries aggrieved at being dragged into the increasingly political row[3]. In Ghana, a political storm erupted over accusations of Airbus payments to a relative of a government official in connection with the purchase of military transport planes. Britain's Serious Fraud Office (SFO) said the agent had no experience in the aerospace industry. Why does such a prestigious company as Airbus lose its sense of value in pursuit of profit?

Many people try legally to avoid paying tax—if they can afford to do so. They may employ tax consultants to get tax efficiency. Any sense of moral or ethical value to pay tax to contribute to the common good is demoted to being a matter of efficiency. Lowering your tax burden by using overseas tax havens may be legal and operating within the letter, but not the spirit of the law or a moral obligation. Systematic tax avoidance by rich people and UK companies feels ugly. But remember, they see this as just a game. And the social norm is that it is a game that anyone with any sense should be playing. It's a game that is so inbuilt that, at the other end of the scale, some people think benefit claimants must be cheating.

But not everyone thinks that way. Many can't because they're too busy struggling to keep their heads above the water. Maximizing expenses and minimizing taxes are luxuries they cannot afford. Most people don't play these games because they can't. But the problems are embedded in our culture.

Individualism of this kind was one of the key factors in the financial crash in 2008 that we will examine in Chapter 5. It's behind a target-driven culture of services like the police and the health service without proper attention to how to reach those targets ethically. It's in parents, trying to make sure they are in the right catchment for the school they want for their children disregarding other important factors. It's in universities trying their best to achieve official research assessment results at the expense of original 'blue skies' imaginative research.

So, what can be done? As I said in Chapter 1, we must change the way we think. The cynic in me says it can't be done. The optimist in me says it can. It requires a common enemy—and we have one—climate change. It requires us to realize that there are some things more valuable than money.

RECOGNIZING VALUE OVER PRICE REQUIRES UNDERSTANDING

To enact change, one needs to understand values and their effects. And that requires evidence. In an address to the Royal Academy of Engineering UK May 2023, the former Government Chief Scientific Advisor Sir Patrick Vallance gave four tips for any scientific advisor. His advice is relevant to us all. (1) Make sure your evidence base is adequate and ask yourself can it be improved. (2) Has the evidence been understood by the people you're talking to? Make sure they understand the uncertainty in the data. (3) Give advice in a form that is relevant to policy makers. (4) Generate and share good quality data.

As regards Sir Patrick's first tip, as I said earlier the evidence base for climate change is established. I don't need to repeat it here. We just need to keep in mind the images of melting ice caps, distressing pictures of flooded homes, scenes of desperate droughts and famine, fish caught up in plastic debris, plankton digesting microplastics and entering the food chain, the extinction of species such as birds in the general decline of biodiversity. Those images will remind us of the dreadfulness of what we face. As to Sir Patrick's second tip it is clear that many non-economists do not understand the uncertainty in-built into policy making regarding our economic health and its consequences for climate change. If we are to put money in its proper place and include natural, human, social, cultural, ethical, and political capital as well as financial capital then collectively every one of us must understand economics better. Of course, we cannot all be experts, but we can attempt to understand at an appropriate level for each one of us individually. The fact is that we cannot just leave it to the specialists. Doing just that has led us into the crises we face in the 21st century. Part of the purpose of this book is an attempt to shed light on Sir Patrick's tips numbers three and four.

A good starting point for most of us is fiat money (first referred to in Chapter 1)—since many of us non-economists have never heard the term before. Yet it is of central importance in understanding the economies of the UK, USA and many other (but not all as will become clear) countries. So, we need to look at it more closely.

You will recall that my learning point number 4 in the Preamble to Part 2 was that fiat money is not well understood by many people in power—especially politicians—and this leads to wrong-headed policy decisions.

What is fiat money? Fiat means by decree. Fiat money is a currency that is declared legal tender by a government that has no intrinsic or fixed value and is not backed by any tangible asset such as gold. In other words, the currency only has value because the government says it does and that government has economic credibility. The country has monetary sovereignty in that it can spend, tax, and borrow its own currency without promising to convert it into gold or foreign currency at a fixed price. The £UK and the US$ are examples. The government *issues* the currency whilst the rest of us *use* it.

If this notion of fiat currency is new to you it may take a while to sink in.

Understanding fiat money enables us to see that the economics of a country that issues fiat money is fundamentally different from the economics of a household or any other users of that currency. The monetary sovereignty that comes with a fiat currency is on a continuum with some countries like the UK, USA, Japan, Australia, Canada, and China having total control. Countries in the Eurozone lack the monetary sovereignty of a fiat currency because they have transferred it to the European Central Bank—the euro € is shared across all member states. Countries like Ecuador and Panama lack sovereignty since their currency is designed around the US$.

Margaret Thatcher is quoted as saying *'Any woman who understands the problems of running a home will be nearer to understanding the problems of running a country'*.[4] She wrongly said that there is no such thing as public money—only taxpayers' money[5]. On both counts she got it the wrong way round. The government issues the money via the banks and recoups much of it back through taxation. Theresa May said the *'government doesn't have a magic money tree'*.[6] Fiat money does provide a magic money tree within certain contextual constraints (mainly real resources and inflation). And of course, it applies only for those governments that issue currency and not for the rest of us. The Euro € is not fiat currency for Greece—only for all the countries in the Eurozone together. When a government borrows its own fiat currency it has monetary sovereignty, and it cannot run out of money. It has its own bank—the central bank and so it can lend to itself. This is quite unlike you and I as individuals and companies that must make sure that their income covers their expenditure—they must balance their accounts. Fiat money is the reason why some government budgets do not have to balance. The surprising consequence is that government debt in those countries is not something that will burden our children and grandchildren (see later learning point 7 in Chapter 6).

One useful way to think about this is via a balance sheet as introduced earlier. A balance sheet is a statement of your assets and liabilities at a point in time (say in April of each year) and the balancing difference is defined as your worth. So, if you own a house that you could sell for £500k and you have a bank balance of £1,000 and you have no other assets then your total assets are £600k. If the money you owe to others such as your bank or mortgage company is £250k including interest payments and you owe other money (have liabilities) of £50k, then your total liability is £300k. That means your net worth is £600k–£300k or £300k. Of course, your actual balance sheet would be much more involved with many more items. You would need to include all assets such as your car etc., and all liabilities such as your utility bills etc.

For a company, the balancing net worth is often called its equity because it is the monetary value of the shares issued by the company. Whilst few of us would normally write out a balance sheet for our personal finances, balance sheets are essential, and required by law, for companies because they provide the information for calculating the rates of return for investor shareholders and for analysts to assess the financial structure of the company. Companies are also legally required to produce statements of income/expenditure and cash flow. Unlike the balance sheet these are written records over a period (usually one year). The income/expenditure statement is what it says—the company's revenues (incomes) and expenses during that period. By subtracting expenses from revenues, we get the company's net income or profit (or loss). A cash flow statement shows how well a company generates cash (liquidity) to pay its debts on time, to fund its ongoing activities, expenses, and investments.

Sector	Non-financial assets	Financial assets	Financial liabilities	Net worth
Corporations*	4.22	27.44	31.00	0.66
Government**	0.86	0.86	3.21	−1.49
Households***	6.09	7.42	2.03	11.49
Total	11.17	35.72	36.24	10.66

* Both financial and non-financial corporations
** Central and local government
*** Households and non-profit institutions serving households

(a)

Sector	Non-financial assets	Financial assets	Financial liabilities	Net worth
Public	1.26	0.88	3.38	−1.24
Private	9.91	34.84	32.86	11.90
Total	11.17	35.72	36.24	10.66

(b)

A simplified version of the detailed balance sheet[7]
(Note that the figures are rounded)

Figure 4.1 UK National Balance Sheet by Asset March 2021—£(UK) bn

THE NATIONAL BALANCE SHEETS

Figure 4.1 is for illustration only. It is a much simplified and overview version of the UK National Balance Sheet as of March 2021[7]. UK national balance sheets are produced annually by the UK ONS (Office for National Statistics) but may initially only be estimates as the collection of data is clearly complex and takes time.

Assets are classified as non-financial and financial. Non-financial assets include land, buildings, equipment, and intellectual property which are not primarily financial but obviously do have financial value. Financial assets are gold, currency and deposits, securities and loans, shares, derivatives, insurance schemes, and accounts receivable (money owed—yet to be received). Liabilities are entirely financial. Figure 4.1(a) shows the national contribution from corporations—financial and non-financial—households and government—central and local. Financial corporations are banks, insurance companies, pension funds and others who deal primarily with financial assets and liabilities. Non-financial corporations are those that primarily produce goods and services and do not, as a principal activity deal in financial assets and liabilities. Figure 4.1(b) shows the relative contributions of the public (government) and private sectors (non-government corporations, and households). Households includes non-profit institutions serving households such as those that provide goods and services, either free or below the market prices, mainly derive their income from grants and donations are not controlled by government. They include charities, trade unions, religious organisations, political parties, and universities.

These financial statements are important not just as a snapshot at a point in time of a company's accounts but because they are essential to understanding the flow of money between all financial entities from individuals to governments (see Appendix: The Flow of Money, for a little more detail on how this works). Unfortunately, financial statements of governments are complex, and require a lot of unpacking. Only the specialists read and understand them in detail. They are mysterious to the rest of us, and we rely on what we are told.

The major government public assets and liabilities (as distinct from private and household) can be grouped in various ways. One way is by identifying fixed assets, financial investments, and working capital. Set against that the liabilities are working capital, financial, pension and those that are long term. UK fixed assets are mainly infrastructure (excluding that which is privately owned such as the energy, water, and telecommunications sectors) including transport (railways and highways), water, land, buildings, and equipment (including the military, and IT systems). Financial assets include student loans, investments of various kinds, foreign currencies, cash and gold and arrangements with the IMF (International Monetary Fund). Working capital includes taxes, money due, prepaid expenses, inventories, and accrued expenditures.

Financial liabilities include bonds and loans of all kinds such as gilts, treasury bills, national savings, and debt such as circulating banks notes and various leases such as the Private Finance Initiative (PFI).

If we are to understand the role of government borrowing then we need to understand the role of bonds, gilts, and treasury bills. Collectively they are called securities. We will look at this in the next chapter in more detail. In the meantime, you will recall from Chapter 2 that the two major parts of national finance are controlled by different agencies. The government sets a fiscal policy. The central bank (in the UK the Bank of England, in the USA the Federal Reserve) sets a monetary policy (see Chapter 5).

In sum, fiscal policy concerns government revenue—chiefly the use of government spending and tax to influence economic conditions.

Fiscal policy changes are political—and take time to enact in a democracy. Fiscal policy contrasts with monetary policy. Monetary policy tries to balance economic growth and inflation. Monetary policy changes can have a more immediate impact. It is important that these two policies work in harmony.

Taxation within fiscal policy is primarily about the government raising, from the public, money to fund public spending. The different ways this is done are called tax instruments or tax structures. But taxation policy can have other objectives too—including reducing inequalities and improving the quality of our lives. Countries may use tax to improve their ability to compete in global markets, but globalisation increases the opportunities for tax avoidance and evasion so international cooperation is vital. Taxes reduce the amounts people can save, reduce the supply of labour and the creation of jobs, as well as a company's levels of investment and innovation. Poor drafting of tax laws can create loopholes for clever economists and accountants to exploit.

Most countries rely on three main sources of tax—income, consumption and other. Income tax is both personal and corporate. Consumption include VAT (Value Added Tax). Other taxes include social security contributions such as National insurance in the UK, estate and import duties. In the last 30 years the proportion of tax revenues to governments from personal income has reduced as top rates have been reduced. The proportion of corporate income tax on profits has increased despite cuts in tax rates and decreased top marginal rate on dividends. Social security contributions have generally increased. Consumption taxes like VAT have declined. The share of property taxes and environment-related taxes has been fairly constant over time. Several countries have introduced tax incentives for investment in research and development. Evidence suggests[8] that recurrent taxes on immoveable property (e.g., buildings) is likely to be the best for growth. Then follows consumption taxes (and other property taxes), personal income taxes and corporate income taxes. This suggests in turn that to achieve economic growth tax reforms should shift emphasis from income taxes to taxes on residential property, but these taxes are particularly unpopular and so few countries do it.

The effects of changes in labour taxes on employment depend on mechanisms for setting wages and the level of minimum wages. A progressive income tax is one where tax rates are higher for higher levels of income. Highly progressive income tax may lower economic growth because for some, it acts as a disincentive for getting a job that pays a higher wage. For instance, some retirees may choose jobs that pay a lower wage to ensure their tax rates don't change. In-work benefits can help low-income households, reduce inequality, and improve efficiency. This is because the gains in levels of employment outweigh the reduced incentives to work as benefits are withdrawn. They help improve returns from up-skilling and reduce the costs of the tax increases needed to finance the benefits.

Reduced corporate tax rates enhance investment and attracts foreign direct investment. The greater the predictability of corporate income taxes the higher are the investments, which in turn, can enhance growth.

Tax policy can also influence productivity. A reduction in top marginal tax rates has been found to raise productivity in industries with a potential for creative enterprise. Lowering corporate tax rates can lead to large productivity gains in firms that are dynamic and profitable. It seems that corporate taxes adversely influence productivity in all firms. This does not seem to apply to young and small firms when they are not very profitable. Multinational enterprises are attracted by tax systems that are stable and predictable, and which are administered in an efficient and transparent manner.

Government Spending

Governments spend their money to improve a county's capacity to supply what their citizen's need. They spend on health and social services, education, and training, national defence, universal credit, state pensions, interest on debt and all other things that support the operation or expansion of national life such as transport. Spending is often classified into three types—mandatory, discretionary and interest on debt. Mandatory spending is based on existing law rather than a budget. It tends to be dominated by health and social security but includes military, housing, food and agriculture and benefit payments. Discretionary spending includes the same areas as mandatory spending but, as the name implies, is spent at the discretion of politicians by setting annual budgets. In the UK, the OBR (independent Office for Budget Responsibility) monitors the public sector finances[9]. They write forecasts for the next five years and compare them with the fiscal targets that the UK Government has set itself. They estimate how much money will be raised from taxes. For example, for 2022–23 UK income from taxation was £987bn and spending was £1,087bn. They estimate how much will be spent on public services, state pensions and interest on debt and whether it will run a budget deficit or surplus (a deficit of £100bn). The deficit is known as public sector borrowing. Income from taxation was 90% (987 * 100/1087) of the total income in 2022–23. When the public sector has a deficit in a particular year it is added to the national debt. But debt can fall as a share of national income if the economy is growing sufficiently. Public sector net debt is the debt minus the small amount of assets that the government could readily turn into cash if needed. One of the UK Governments fiscal targets is to ensure that debt is falling relative to national income by 2025 to reverse the big increase in spending from the financial crises and pandemics. The OBR report that the UK government raises less revenue relative to national income than most other industrial countries—more than the US, Japan, and Korea, but less than Scandinavian countries like Denmark and Norway. Public spending as a share of national income in the UK is slightly above the average of other industrial countries—the UK spends much more than Japan and Korea, but much less than Finland or France. Spending is just above the international average, but because it raises less revenue, the UK runs a bigger budget deficit than the world average. Net debt in the UK is also higher than the average of other industrial countries.

The role of deficits and net debt in countries with a fiat currency is a subject of important and controversial debate amongst economists. We will return to this in Chapter 5 because it has a direct impact on investment in managing climate change. Meanwhile we note that the lack of a settled view does not help politicians to make decisions and the public to decide to vote for when deciding what are the best economic policies.

Nonetheless whatever the individual policies are it is obviously crucially important that these two policies, government fiscal and central bank monetary, work in harmony.

STATISTICS IN ECONOMICS—ECONOMETRICS

Many academic economists think that mathematical and statistical modelling is essential to understand value and price. That is how they think we ensure rigour. Econometrics is the branch of economics concerned doing just that—to quantify economic phenomena. As we have already highlighted economic theory contains many models of understanding of the relationships between important measures of economic performance. In Chapter 5, we will look at an important model devised by David Li[10] that played a crucial role in the banking collapse of 2008. His model uses the Gross Domestic Product or GDP first mentioned in Chapter 1.

In simple mathematical terms (GDP) = Consumption (C) + Investment (I) + Government spending (G) + Trade balance (NX). Relationships like this can be set down in mathematical form—in this case just simple addition of economic variables. Again, if specific numbers are found (by survey[11]) for each of the variables, C, I, G and NX then there is an implied assumption that they are deterministic. Of course, they are not and most of us realise that. But if the measures or variables are not deterministic then what are they? Most statisticians assume that the next best thing is to quote a range of possible values such as GDP = £987bn ± £2bn. The problem for the statistician is then is how to determine what the range should be. The most common way is to model the variables as so-called random variables[12]. These are not 'truly' random. Truly random numbers have no detectable or specific order. For example, a 'truly' random sequence of numbers such as 9, 3, 5, 4, 10 has no detectable pattern. Another example is the tossing of a dice where there is an equal chance that each one of the 6 faces will turn up. Random variables in a statistical model are chosen 'truly' randomly but taken from an underlying distribution of a frequency of occurrence in a set of possible numbers. The most well-known example of this is the bell shape or normal distribution[13]. Now the question is how are these numbers chosen and what are the uncertainties in the resulting values?

Sampling

If we wanted to find the distribution of the heights of all people living in the UK then of course, it is impractical to measure everyone in the total population. Likewise, it's impossible to ask every TV viewer what they watched on TV the previous evening. That is why samples are used. A random sample is one where the members of the sample are chosen with an equal chance of selection[14].

But there are always sources of bias due to approximations, people refusing to participate, flaws in measurement instruments as well as straight human errors.

We can classify the errors into three types—random, systemic, and human (often called gross) errors. Most often if the bounds on an error are quoted by statisticians, then they are based on analyses of random errors. These result from random fluctuations in any set of measurements. For example, if you measure one person's height very accurately many times then each time you will get a slightly different answer. The more accurate you try to be the greater will be the apparent variations. If you plot a graph of the variations, then you will get a normal distribution. We would usually take the mean of that distribution as our best answer. However, if we also note the spread, we can quote a range of possible answers between say –2 and +2 standard deviations[15]. Where you find an error bound quoted it will almost certainly be based on that kind of analysis of random sample data. We have already noted that TV viewing numbers are usually quoted as a single number with no range or distribution.

Systemic errors are errors are more difficult to spot. These are errors of methodology and technique. Examples might be measuring instruments (like weighing scales) that are not calibrated correctly or are deficient in some way, methods of data collection and analysis that are not fit for purpose, plus other effects like an increasing rate of refusal from people unwilling to answer survey questions. One of the reasons why systemic errors are very difficult to estimate is because they often go unnoticed. That is why it is good for people to work in teams to pool ideas, check calculations and identify and eliminate systemic errors as far as possible.

Human (or gross) errors are also difficult to identify. They can be individual mistakes (as simple as adding up numbers incorrectly) or collective cultural biases (as complex as phenomena such as groupthink[16]).

One of the central problems for inferring conclusions from statistical data is managing these kinds of error—removing them where possible, allowing for them where necessary. Modern complexity theory[17] has demonstrated to us the importance of resilience in the face of unintended consequences when unexpected events occur. When collecting sample data how do we allow for the possibility of unknown unknowns—things you don't know that you don't know[18]. No sampling before the financial banking collapse of 2008 included the possibility of that event—though some people claim to have spotted the signs in the data. A major lesson that history tells us is the need for transparency. The methods used to sample, collect, and draw conclusions from data should be clearly described and a balance drawn between various quality characteristics including relevance, accuracy, timeliness, accessibility, interpretability, and consistency as attributes of fitness for purpose.

All too often such clarity is not available, and it is rare in economic performance data. So how are we to know what is happening in the economy? That is the main purpose of this chapter. How do we measure economic activity? How do we know what are the important factors that affect our decisions? Whether we are irrationally rational or not, how do we use the enormous amounts of data that is available to us and how do we interpret it?

Data and Information are not the Same

We are often told that we live in a data driven society now. Data is used for more efficient, decentralized decision making. Modern IT produces masses amounts of data from a whole range of sources. It is common to hear data described as a collection of raw numbers, statements, symbols, and facts. The word data can be used for a singular fact or a collection of facts. It comes from the Latin word datum, meaning "something given". The word datum is still the technically correct singular form of data but is rarely used in common language. At its simplest data are simply facts and figures. In the world of computers data is the input. Data is what you tell the computer to do. In the world of statistical analysis, data is raw information, but the statistical analyses help us to summarize and interpret the data. In the world of business, data are also usually raw numbers with individual data points that help us understand what has been measured.

But data is not information. At its simplest information is news or knowledge received or given. It is what results when you have processed, interpreted, and organized data and facts. The word comes from the Latin word *īnfōrmātiō*, meaning formation or conception. Information is how we understand data. But this can be seriously misleading when 'so-called facts' are disputed. To see why we need to look more deeply at the difference between data and information. We need to examine the role of measurement, the nature of evidence and how we tell the difference between that which we can depend on and that we cannot.

Let's start by recalling Daniel Kahneman and Lisa Feldman Barrett's work mentioned in Chapter 3 thinking when we were looking at rationality and irrationality and patterns of connections in our brains. We can regard data as patterns of our five senses, sight, sound, touch, taste, and smell that most of us have. We are receiving sensory patterns all the time through our experiences of the physical world. Even as we sleep, we are receiving sensory data and processing it in our brains and bodies. In our social interactions with others sensory patterns are transmitted from our sense organs to our brains. There they are translated and formed into patterns of connections between our brain cells or neurons. In turn the nucleus of our nervous system, the brain, is informing patterns of cells in our bodies enabling us to think and act in our physical and social environment. Information is the meaning that we impart in our brains to those patterns some of it conscious, as in our awareness and ability to think, and some of it in our subconscious such as controlling heartbeats and our digestive functions[19]. Through our eyes we detect (see) patterns of shape and interpret them. As babies they are relatively meaningless but as we grow and learn we begin to make associations between patterns. For example, we see shapes and learn that they are those of a cow or a cat. We detect (hear) patterns of sounds and learn to interpret them as a moo or a meow. We touch a rough surface and learn to interpret it as a stone or brick. We smell and taste a fine wine and enjoy the sensation. We see symbols on the page of a book as letters and we learn to read them. As you read you recognize patterns of letters as words and patterns of words as sentences. You can confirm this by looking at a foreign language script such as Sinhala and realise that it means nothing to you—but to someone born and raised in Sri Lanka it has meaning.

Measurement

Measurement is a process in which two observable phenomena are compared. One is a known or standard phenomenon, and the other is not. For example, when we measure the height of a person, we compare that person's height with a standard unit of defined length such as a metre or a centimeter (or in old language feet and inches). The dependability of the answer we get depends on four sufficient conditions on this process[20]. First the process must be repeatable, for example, in classical physics we can measure something over and over again knowing that the system has not significantly changed. Second the state of the system must be repeatable, that is when we can repeat a measurement, we need to be sure that the state of the system is the same each time and not changed as is possible in the unstable buckling of a slender strut. Third the state of the system must be very clearly defined, not fuzzy, ambiguous, or imprecise—a dirty tape with faded printing could be difficult to read. Fourth the magnitude of the measure must be acceptably close each time, for example, the length of the line is approximately the same each time we measure it, albeit with random variation. For most everyday measurements all these four conditions are satisfied and taken for granted. Measurements of samples of statistical data do not always meet these conditions. For example, when you throw a dice the state of the outcome is the face of the dice and that may well be different for different throws. If you measure the buckling strength of a long rod, then the rod may buckle in different ways resulting in different states. The state bifurcates due to instability—a phenomenon at the heart of complexity as we shall see. Where the results of a measurement interact with the observer the state of the process may change from what it otherwise may be. Measurements in the social sciences are notoriously difficult because it can be difficult if not impossible to meet the four requirements sufficiently well. For example, measurements of opinions or confidence levels in ballots are achieved by voting processes. In economic utility theory measures are found by betting on outcomes. But the answers we get from these kinds of processes and surveys depend crucially on how the questions are phrased. You may recall that these are systemic errors.

The first requirement for any measurement is to define and clearly communicate the purpose of the process we will use to take the measurement. The listening and observing and the collective understanding that follows can create added value, so the measures are acceptable if the process and results are judged appropriate and fit for purpose. When we measure a judgement, we must ascribe a worth to something. Money as defined earlier FEW is one very important way of doing that. But it is not the only way as demonstrated by my earlier introduction of the obligations of TWO. The worth of softer values is partly personal and partly shared. We must recognize therefore that the answers may not be as dependable as we might wish. That does not mean that the results cannot be useful if we are very clear about the purpose of a measurement. The purpose and process of measurement is as important as the result.

Evidence

Evidence is usually described as 'something that is true or leads to a proof or conclusion'. As I have said, one of the illusions of life is that we can establish what is true in all contexts—find the universal truth. Philosophers have struggled with this notion for millennia. Of course, we can believe that a statement is true but proving that to everyone else's satisfaction is elusive. Remember from Chapter 1 that common sense tells us that if we are holding a stick and we let go it will fall to the floor—gravity. But that is not always the case as Edward de Bono pointed out[21]. If we are under water the stick may rise to the surface and if we are in a space capsule it will stay where it is and float in space. The statement we make depend on context and that context may not always be clearly stated or even understood. A better way of thinking about how we use information to decide, and act is to look closely at the dependability of the evidence.

The dependability of a piece of information relies on how well it has been tested. Fake news and conspiracy theories spread when they are not sufficiently tested in context. Fake news is not dependable. Testing does not imply that all contradictions and tradeoffs will have been eliminated. Scientists know that the best way to test is to measure that which can be measured. But often it is impossible to meet the four conditions of a measurement process. Underpinning the dependability of evidence is critical thinking and critical discussion. Critical thinking is the ability to question information, ideas, and arguments. It is about asking the right questions, weighing up different arguments and perspectives and synthesize the evidence to help form well-reasoned arguments that can be justified to others. Critical thinking is about questioning and learning with an open mind. Differences of opinion will inevitably emerge about incomplete and complex matters but wherever possible we want to know it has been tested in reality. So, when a structural engineer designs a pile of steel girders to make a bridge, we know he/she is using well tested theories and experience. When an economist recommends a particular policy, we should be able to assume that the theories on which the policy is based has been tested by previous application in the past or by thorough critical discussion. Therein lies the difficulty. Testing financial and economic models that are fit for purpose is not straightforward.

Correlation is not Causation

The way we learn to associate, recognize, and interpret patterns to impart meaning is central to how we understand and behave in the world around us. Despite the apparent simplicity of everyday experiences, such as pouring a drink or reading a newspaper, the way we learn to cope with life contains a hidden complexity. That complexity comes to the fore when we are faced with very difficult decisions.

The way we typically learn is through realizing that there are constant conjunctions between brain patterns. But those conjunctions are not infallible. The philosopher David Hume (1711–1776) is well known for pointing out that they can't be relied upon. A constant conjunction is a relationship between two events in which one event is invariably followed by the other. In philosophy this is known as the problem of induction—how we reason from the particular to the general. For example, if every swan we have ever seen in our whole lives so far is white, are we justified in asserting that all swans are white? The answer is no we are not. The reason is that there are, perhaps unbeknown to us, black swans. Our thinking depends on the width and depth of our experiences. The more we know the more we realise how much we don't know. Fixed minds often derive from limited exposure to all that life has to offer. Of course, we also learn about the experiences of others as well as ourselves through storytelling and discussion—a form of social interaction. But there are inevitably things we don't know as well as things we don't know we don't know. This is just a fact of life.

Correlation and causation are just two of the words commonly used to describe constant conjunctions. But they are very different ideas. Both express a relation between two or more things, the degree to which two or more attributes or measurements on the same group of elements show a tendency to vary together. But correlation is not causation. Causation is when one event causes another to happen—cause and effect. For example, if you exercise you cause your muscles to develop. Rainclouds cause rainfall. Dropping a wine glass caused it to break. The two types of cause are those that are *necessary* and those that are *sufficient*. A necessary cause must happen for the effect to occur—the effect cannot occur without it. A sufficient cause is one that is enough for an effect to occur—the cause is all that is required for the effect, but the effect can occur from other causes as well.

Correlation is constant conjunction without causality. For example, every time you toss a die with your fingers crossed you get the number you predicted. The crossing of your fingers does not cause the dice to fall in any specific way—nevertheless that is what seems to happen. It can border on superstition.

In mathematical statistics correlation has a slightly different and narrower meaning. It captures the idea of a linear relation between two variables. That means that as one variable changes, the other changes in proportion. If you plot those changes on a graph, then they show up as a straight line and not a curve. The equation for GDP we saw in the previous section is an example of a linear equation. A circle drawn on a 2D graph of variables x and y is a clear relationship but the correlation coefficient between x and y turns out to be zero.

Logic, Induction, Deduction, and Abduction

In Chapter 3 we looked at rational and irrational thinking. Now we are beginning to see how difficult it is to be rational. Rational means being logical. Logic is about how to infer or make dependable conclusions from a set of premises. Logic underpins mathematics and is itself subject to controversies. The three forms of logic that we described in Chapter 3 are induction, deduction, and abduction. We rely on them to interpret data to make sense of what we are told and how we infer meaning. You will recall that an inductive logic is a logic of evidential support where, as we said earlier, we reason from lots of specific cases to more general assertions—from saying that every swan we have seen so far is white to all swans are white. Deduction is a stronger form of reasoning, but it depends on some premises that are assumed to be true. Mathematics is a form of deductive reason based on sets of premises called axioms. In a deductive logic, the axioms logically entail the conclusion. That means that every possible situation that makes the axioms true must also make the conclusion true. The axioms provide total support for the conclusion.

Abductive logic is where all we can say is that the axioms provide probable support for a conclusion—it is the weakest from of logical argument.

One way in which induction is implied is called case-based reasoning. This is where new problems are solved by referring to lessons from similar problems of the past. For example, a car mechanic faced with an engine that won't start may realise what the problem is by remembering another car that had similar symptoms. Whilst inductive reasoning can be effective in this way when faced with complex issues such as climate change it may not work because there are few, if any, precedents to go on.

Complexity

Complex systems are deeply entangled. Everything seems, and often is, connected to everything else. Complexity makes making sense of data much more difficult. Complexity can mask the distinction between price and value, for example, distinguishing between the price and the cost of delivery of very some complex goods and services like airports and other major infrastructure projects. First estimates of cost are often nothing like the final cost because the procurement of an airport is so complex. In the 21st century we have begun to realise that the physical world as well as the social world are inherently complex. Because of the masses of patterns thrown at our senses and our brains daily we often make conscious decisions based only a partial scan of the patterns we have learned. Some people do this more quickly and readily than others. Consequently, we tend to see what we expect to see rather than what is there. We pick up on the patterns we have learned rather than puzzling over new ones. Indeed, we tend to turn the patterns into stories or narratives that appeal to us. We get caught up in the hurley-burley of similar stories—tropes that attract us and form, you will recall from Chapter 1, our 'worldview'. Each of us develops our own worldviews—perspectives and views on life—that are similar but different. Groups brought up in similar traditions have similar worldviews although individuals within those groups have specific views that are entirely their own. Scientists think in ways that are shared within the group, as do engineers, medics, statisticians, economists, and theatre actors. Each individual scientist, engineer, medic, statistician, economist, and theatre actor has a similar way of thinking but interpreted slightly differently from others in the group. But the worldviews of different groups may be dramatically different. For example, statisticians and scientists will want to see mathematical models. Theatre actors, in general, do not find mathematical models at all useful in deciding how to interpret a role in a film or play.

One way of thinking about group culture is the day-to-day stories that they tell each other and the rest of the world. Those stories are often reinforced by daily experiences—quite often uncritically. Complexity contains causalities that are not linear as discussed earlier. It contains many bifurcations from instabilities in the developing state of affairs within a system. The consequence is that effects from causes may be quite delayed in time and hence difficult to associate. Deciding on specific actions to produce effects may change the causal chains of influence. For example, targets can be self-defeating when they distort behaviors because people find ways of working around them. When UK GPs were set a target of seeing patients within 48 hours the consequences were that practices only made appointments for two days ahead and many patients could not get to see their GP within a reasonable time.

ECONOMIC INDICATORS

What does this all mean for economics and climate change? Most importantly we must treat economic systems as complex. There are at least five reasons for this. First, as I have said, cause and effect may often be difficult to identify and delayed in time and not linearly related. Second, the patterns of performance and behaviors in us, and in the physical, natural world, are perhaps best understood as massive networks of communications between many different holons. You will recall from Chapter 1 that a holon is at the same time both a whole and a part. Third the behavior of a 'whole' may be more than the sum of its parts. The 'whole' may have properties that emerge from the interacting parts. Fourth, systems can bifurcate or flip from one state to another quite suddenly—as for example during the financial crash of 2008. Fifth, future outcomes are highly uncertain. Deterministic models are no longer sufficient for all but the simplest of systems. Even stochastic models that rely on random variables and seem to predict chances of various future states are inadequate. This realization that unanticipated and unwanted sudden shocks can result in massive change in a very short space of time has only come about in the 21st century. We are beginning to appreciate the importance of guarding against the unintended consequences from the decisions we take and the need to be resilient when unknown unknowns cause huge difficulties and possibly massive damage and loss of life.

There are two responses, at least, to these complexities—one pragmatic and the other perhaps rather idealistic, and some will say naive. The pragmatic approach relies on individual judgements based on education, training and experience of decision making in practical situations. For that we need to nurture practical wisdom as described in Chapter 3 and elsewhere in this book. The other is to build ever more elaborate mathematical models. Both approaches have strengths and weaknesses. The pragmatic approach is abductive. It evaluates theories or beliefs from practical application—the main concern is getting practical consequences from decisions and actions. The problems are many but chiefly involve the uncertainties of applying past experiences to future situations. Idealism, in its purest form asserts that reality is best understood through thoughts and ideas. The advantage is that it can be very general and, at the same time very specific in its predictions of what the future may hold. Its disadvantages are that it can become very difficult for anyone other than specialists to understand, mathematically complex and subject to significant uncertainties about its relationship with reality.

Economic indicators are pragmatic macroeconomic measurements. They are partial glimpses of a complex whole rather like looking at the moon and seeing a disc when we know it really is more like a spheroid—a squashed sphere. What we see is only a partial glimpse of the total moon. We are condemned to piecing lots of partial glimpses from different angles together, to form a picture of the whole. How can we know that the moon is a spheroid when all we see is a circle?

Governments[22], non-profit making organisations and universities publish and pore over national indicators. Company directors employ staff to produce data, information, and indicators for the Board of the company to interpret. The Board use these data to make decisions about the company and they are required by law to publish certain data as financial reports[23] usually within four months of the end of a financial year[24]. Each of us as individuals may take an interest in some indicators that may affect our personal financial health.

Economists divide indicators into those that are leading or lagging. Leading indicators tend to precede or predict trends. Lagging indicators tend to confirm or deny trends and reflect what is happening now. Some economists also identify lagging indicators as coincident[25] indicators. GDP, employment levels, and retail sales, are coincident indicators as metrics of the activities within a particular area or region. Other lagging indicators are gross national product (GNP), the consumer price index (CPI), and unemployment. Leading indicators include interest rates, stock market indices, such as yield curves[26], consumer durables[27] house prices, retail sales, and net business formations[28].

The first indicator amongst equals is the GDP (Gross Domestic Product). It is the most quoted measure of the financial value of the production of goods and services in a country during a certain period and yet it is widely recognized as inadequate.

One of the reasons is that GDP includes costs and waste, i.e., activities that may be unproductive or even destructive. Although, it includes the income earned from production, or the total amount spent on final goods and services (less imports) it excludes informal economic activity such as unremunerated volunteer work, leisure time or household production, profits of foreign investors and activities between businesses. In the context of the increasing urgency of addressing climate change and its effects many people are calling for the use of different measures of economic performance. We will look at that in more detail in Chapter 7.

Stock market indices such as the FTSE 100[29], Dow Jones Industrial Average[30] and the S & P 500[31] are a good predictive indicator of economic health, because traders spend time assessing the potential for growth. But the drawback is that prices are speculative and can become over and under-valued. The UK house price index is published by the ONS (Office for National Statistics) every month[32]. A declining house price index suggests that the number of houses for sale is greater than the number of potential buyers. The impact on the wealth of a homeowner wealth, and on jobs in the construction sector may force some homeowners to foreclose where lenders try to recover their mortgage loans. As we will see in Chapter 5 through the 2008 subprime financial crisis the housing market can impact the whole economy. The number of applications for building permits also indicate economic health because companies will apply for these permits at least six months before they start construction—an indication that they expect demand for homes to rise. Bond yields indicate the income that a trader can expect to receive in return for buying and holding a bond. They are published daily for government bonds. Short term bond yields are directly impacted by central bank decisions and expectations about interest rates whereas longer-term bonds are affected by interest rates, inflation, and economic growth. Changes in bond yields are often used to examine the economic outlook. Uncertainty flattens any growth in yields because investors become indifferent to maturities. When short-term bonds yield more than long-term bonds investors may be expecting economic growth to slow down and central banks to cut interest rates. One indicator of production and manufacturing output is the Industrial Production Index (IPI) published monthly[33]. It shows the growth of various sectors in an economy such as mineral mining, electricity, and manufacturing. It is often a composite measure of the short-term changes in the volume of production of a basket of industrial products during a given period. Increasing levels of production and manufacturing outputs tend to increase GDP and is taken as a sign of increased consumption. Production levels also affect employment rates. The number of available jobs in manufacturing shows how confident businesses are in their ability to expand perhaps showing they have a full order book. If inventory and retail levels are up, then consumer demand may be up unless there is stock piling in warehouses. Labour productivity is also measured as output per hour worked and published quarterly by the UK ONS (Office for National Statistics). The Retail Sales Index is published monthly. Levels of retail sales of goods and services by consumers and business indicate strong spending by consumers and the economy is improving. Levels of manufacturing will rise commensurately and boost GDP. Share prices may also rise. When consumers start to feel the pinch, they stop buying unnecessary items and restrict their spending. The downside is that retail sales don't necessarily provide an accurate picture because people may be borrowing to spend, and rising consumer debt levels may indicate an impending recession. Interest rates can be argued as being both leading and lagging indicators. They are lagging when decisions by central banks are reacting to an economic event or market movement. They are leading because the economy will change in response to the new rate. When consumer spending and inflation rates are both high central banks will raise interest rates reduce the pace of growth. The higher rates mean that banks will have to pay more to obtain money and that will increase the cost of borrowing for consumers. Consumers will become reluctant to borrow money and reduce spending. If the economy is stagnant central banks will lower interest rates to boost spending, so that the cost of borrowing will go down, spending will increase, and the economy will start to grow.

When GDP increases, other indicators such as employment rates will grow for example as companies take on more employees and increase production. A consistent GDP rate of growth is taken to be a good sign that the economy is stable but there are issues as we will see in Chapter 7. Some say that it is too easy to manipulate GDP figures. For example, up to 2019, India seemed to be the fastest-growing major economy with annual GDP rates of 7%. Then a flaw in the measuring process was found and the rate was lowered to 4.5%.

If the GDP rate declines two quarters in a row, then the economy is entering a downturn or recession. Inflation is a lagging indicator of economic growth. A high rate severely impacts the price of a currency, making products more expensive possibly leading to unemployment, lower growth of GDP, and rising interest rates as attempts to control prices. Deflation is negative inflation with reduced consumer spending, retail sales, money supply, and rising unemployment rates. The USDX (US Dollar Index or Dixie) is an index of the value of the US$ relative to a basket of other currencies, including the £GB, Euro, and Japanese Yen. The strength of a country's currency is a lagging indicator that reflects the political and economic health and stability of an economy. Significant uncertainty or change can cause rapid changes or volatility. The healthier an economy, in the view of investors, the more they will pay for a currency. A strong currency has higher purchasing power. If a country is a net exporter increased currency prices means that goods can be sold at higher foreign prices, but importers might not want to pay these increased prices. If a country is a net importer, foreign goods become cheaper. Investors may not want to invest in a weak economy, so the currency declines in value, lowering the price of exports and making them more competitive globally. A weak currency may encourage tourism and demand for domestic goods.

Two of the most useful lagging indicators are the rate of employment/unemployment published monthly in the UK by the ONS (Office for National Statistics)[34]. Increasing unemployment over a period, indicates that the health of the economy is declining, and businesses are laying off their workers. Even when the economy is starting to improve, employment rates might not immediately recover since employers will wait until they are sure their business and the general economy is growing. A Commodity Price Index is a weighted average of a selected commodity prices. They tend to change before other lagging indicators. An increase in demand for commodities, such as wood, steel, and energy, are a sign that an economy is growing because they are needed for building infrastructure. The price of some commodities, such as gold, will increase during economic downturns because gold is a safe haven. If the price of gold declines, then investors may be moving their money into higher-risk assets.

As we have stressed all these indicators have to be interpreted with a 'pinch of salt' since they are highly uncertain and may even sometimes be plain wrong. Economists rely on a wide use of ratios to interpret large and complex financial performance.

It is all too easy to underestimate the contextual nature of such data. Interpretation requires judgement and experience. A ratio is perhaps the simplest form of a bilateral mathematical relation[35]. Mathematical functions (linear and non-linear) express relations between many variables. Given that econometrics is the application of statistical and mathematical models of the economy then the widespread use of models such as linear regression[36] may mislead.

Artificial Intelligence (AI) may be used to trade stocks and inform decisions about derivative trading. In all cases interpretation is tricky. It requires a good dose of practical wisdom[37] which all too often is sadly lacking. Some of these ratios are thought to be roughly constant in the long run and adopted in models of balanced growth, arbitrage,[38] or solvency.

Other ratios used include working capital ratio[39], the quick ratio[40], earnings per share[41], price-earnings[42], debt-to-equity [43], and return on equity[44].

Relying on Indicators

Clearly it is not easy to interpret a whole set of changing indicators to understand and control the economy. But experts can and do because they must. Because the system is complex opinions and judgements vary and the understanding, especially of politicians can be more or less seriously incomplete.

An example of how economists' reason is as follows. If we increase the supply of money[45] then this should lower interest rates and hence generate more investment[46]. That should put more money into the pockets of consumers and stimulate more spending. More spending by consumers helps businesses to grow as they buy more raw materials increase production and recruit more labour. If the money supply falls the opposite occurs.

Increased spending by governments[47] can cause a rise in overall demand for goods and services but that can also lead to inflation if the supply of those goods and services does not meet that higher demand. Governments spend for several reasons including welfare, reducing inequalities, educating, and training, providing goods and services, improving infrastructure—all with the aim of increasing economic growth. But there is a downside. The spending can sometimes lead to inflation, higher borrowing, higher taxes, and higher bond yields. If governments wish to spend to help increase supplies to meet demand, then it must be focused. For example, governments can improve infrastructure through better transport and communication networks, by removing supply bottlenecks and hopefully enabling greater efficiency and improved productivity. Government spending on welfare benefits and pensions will produce different effects. They spend to help people in extreme poverty, suffering from ill health or disabilities. They also spend to reduce inequalities. The risk is that the spending may reduce investment by the private sector. Increased welfare benefits help the unemployed and reduce the risks of extreme poverty but could in some circumstances reduce incentives to get back to work or retrain. Government spending by increasing state pensions helps an ageing population but has little effect on productivity. Spending on education and training can improve skills and increase productivity.

Government spending can increase annual deficits and cumulative national debt. Higher debt may result in increased interest rates and higher debt interest payments. In other words, borrowing costs more. Government spending also depends on how it is financed. Most government revenue (typically 90%) derives from taxation. If spending is financed by higher taxes, then people will have less to spend. The impact of government spending depends on the state of the economy. When an economy is close to full capacity increased government spending may be inflationary with little increase in real GDP. It may 'crowd out'[48] private sector spending and investment because savings are lower. On the other hand, if an economy is in recession, and the government borrows from the private sector then an expansionary fiscal policy can boost economic growth. Some economists argue that government spending is more inefficient than private spending through poor incentives and misallocation of resources.

As interest rates are raised the costs of borrowing costs also rise for everyone and people eventually start to spend less. The demand for goods and services drops, and inflation will fall. Of course, there is always a chance that some borrower will not repay the money and so interest provides a degree of protection. Lower interest rates result in more people willing to borrow money to make big purchases, such as houses or cars. There is also a chance of a 'ripple effect' of increased spending throughout the economy. The output multiplier[49] expresses how many £UK of increased economic activity result from a £UK reduction in taxes or increase in government spending. Businesses and farmers may make large equipment purchases so output and productivity increases. High levels of interest rates reduce the disposable income of consumers and hence they spend less, and banks make fewer loans.

Tax cuts increase the disposable income of consumers, boosts demand and encourages businesses to invest more in labour and equipment. Clearly tax increases do the opposite. These effects on demand effects can be substantial when the economy is weak but are smaller when it is operating at near capacity.

Uncertainty

The world is becoming a riskier place. Uncertainty is high. This brings with it extra costs, many alternative views and fundamental disagreements. Fiscal support for households, firms, and public services during the Covid-19 pandemic was unprecedented in scale, reaching 10.4 per cent of GDP at its peak in the UK. As I write this the UK Government is topping up that spend in 2022 onwards to meet the sharp rise in the cost of energy (gas and oil) and hence the cost of living due to the Russian invasion of Ukraine. Uncertainty is fueled by, and a consequence of, successive shocks. The UK Government's response to them has resulted in public debt levels of over 90 per cent of GDP, and more than triple its level at the start of the century and more than double the around 40 per cent of GDP projected by the Treasury in the UK's first and pioneering long-term public finance report published two decades ago[50]. The UK has also felt the economic and fiscal consequences of the uncertainty created by the UK's decision to leave the EU and the ensuing negotiations. Governments around the world face the even greater economic and fiscal challenges of addressing climate change, dealing with the fiscal costs of ageing, rising levels of mental distress, political polarization, and rejection of democratic norms. They must manage all these pressures and risks against a backdrop of potentially weaker productivity growth, higher levels of public debt, and rising interest rates, cyber-attacks from all sources, fragmentation of international financial systems, and unknown future shocks. But we have to learn to do it across national boundaries without killing each other. Wars are a distraction we could do without.

You don't know what you have got until its gone. The Covid-19 pandemic was a reminder, if we needed one, that the everyday things we take for granted can be suddenly snatched away. Recognizing value over price requires understanding. If we are to put money in its proper place and include natural, human, social, cultural, ethical, and political capital as well as financial capital then collectively we must understand economics better and appreciate its uncertainty. We cannot just leave it to the specialists. One of the most important misunderstandings by eminent politicians from Margaret Thatcher to Barack Obama and many currently in power is the role of fiat money and these have led to wrong-headed policy decisions. Governments that issue fiat currency can operate differently from those of us who use it. Few of us can read the national accounts even though they are published. They are very difficult for the layperson to read and interpret. Simpler versions should be available—for example of balance sheets because they are essential to following the flow of money between all financial entities from individuals to governments. Fiscal policy concerns government revenue—chiefly the use of government spending and tax to influence economic conditions. Monetary policy is the control of the supply of money by a central bank by altering interest rates and by buying/selling government bonds.

Governments spend money (as investment) to improve a county's capacity to supply what their citizen's need. They spend on health and social services, education, and training, national defence, universal credit, state pensions, interest on debt and all other things that support the operation or expansion of national life such as transport. We are aware of national deficits and accumulated debt but how many of us have any idea of a country's net worth. Rather than profit we should be looking at net worth to assess financial value.

Cost-benefit analysis is a common way to assess decisions and actions that largely ignores wider values. It requires measurable financial metrics such as revenue earned, or costs saved. But how do we value, for example, the cost of a human life, a loss of biodiversity or pollution of the air we breathe and the water we drink.

How do we tell the difference between that which we can depend on and that we cannot? Econometrics is the branch of economics concerned with the use of economic theory, mathematics, and statistical inference to quantify economic phenomena. Economic theory contains many models of understanding of the relationships between important measures of economic performance. Samples are widely used. A random sample is one where the members of the sample are chosen with an equal chance of selection. But there are always sources of bias due to approximations, people refusing to participate, flaws in measurement instruments as well as straight human errors. We can classify the errors into three types—random, systemic, and human/gross errors.

Data and information are not the same. Data is raw and often is a collection of numbers, statements, symbols, and facts. Statistical analyses help us to summarize and interpret the data. But data is not information. Information is news or knowledge received or given and measurement is a crucial part of it. Measurement is a process in which two observable phenomena are compared. One is a known or standard phenomenon and the other is not. Evidence is usually described as 'something that is true or leads to a proof or conclusion'. One of the illusions of life is that we can establish what is true in all contexts—that is universally true. The way we learn to associate, recognize, and interpret patterns to impart meaning is central to how we understand and behave in the world around us. Despite the apparent simplicity of everyday experiences, such as pouring a drink or reading a newspaper, the way we learn to cope with life contains a hidden complexity. That complexity comes to the fore when we are faced with very difficult decisions. As we saw in Chapter 2, the way we typically learn is through realizing that there are constant conjunctions between brain patterns. But those conjunctions are not infallible. Rational means being logical. Logic is about how to infer or make dependable conclusions from a set of premises. Logic underpins mathematics and is itself subject to controversies. The three forms of logic are induction, deduction, and abduction. An inductive logic is a logic of evidential support where we reason from lots of specific cases to more general assertions. Deduction is a stronger form of reasoning, but it depends on some premises that are assumed to be true. Mathematics is a form of deductive reasoning based on sets of premises called axioms. In a deductive logic, the axioms logically entail the conclusion. That means that every possible situation that makes the axioms true must also make the conclusion true. The axioms provide total support for the conclusion. Abductive logic is where all we can say is that the axioms provide probable support for a conclusion—it is the weakest form of logical argument, but it is the most practical.

Complex systems are deeply entangled in that everything seems to be connected to everything else. In the 21st century we have begun to realise that the physical world as well as the social world are inherently complex and uncertain. It follows that we must treat economic systems as inherently complex. There are at least five reasons for this. First cause and effect may often be difficult to identify and delayed in time and not linearly related. Second, the patterns of performance and behaviors in us, and in the physical, natural world, are perhaps best understood as massive networks of communications between many different holons. You will recall that a holon is at the same time both a whole and a part. Third the behavior of a 'whole' may be more than the sum of its parts. The 'whole' may have properties that emerge from the interacting parts. Fourth, systems can bifurcate or flip from one state to another quite suddenly—as for example during the financial crash of 2008. Fifth, future outcomes are highly uncertain. Deterministic models are no longer sufficient for all but the simplest of systems.

If we are to switch the narrative from liberal capitalism to principled capitalism and from the consumer story to the citizen story, then we need to delve into the complexity of monetary theory and economic policy much more deeply than we have so far. In doing so we can perhaps begin to correct some of the misunderstandings that lead to bad economic decisions—both fiscal and monetary, and we can better hold our politicians to account when they display their lack of understanding. That is what we must now turn to in Chapter 5.

End Notes

1. Oscar Wilde in Lady Windemere's Fan has Cecil Graham ask Lord Darlington 'What is a cynic? The reply was '*A man who knows the price of everything and the value of nothing'*. See https://www.freebooknotes.com/quotes/lady-windermeres-fan/
2. Mayer, M., H. Mott, A. Henderson, C. Marren and A. Bazeley. 2020. Tackling sexual harassment in the workplace. Report on employer actions to prevent and respond to workplace sexual harassment. https://www.fawcettsociety.org.uk/news/over-40-of-women-experience-sexual-harassment-in-their-working-lives
3. The bribery scandal reverberated around the world as investigations were launched in countries aggrieved at being dragged into the increasingly political row. (Reuters, 2020) https://www.reuters.com/article/us-airbus-probe/airbus-bribery-scandal-triggers-new-probes-worldwide-idUSKBN1ZX2MW
4. Margaret Thatcher wrongly said that there is no such thing as public money—only taxpayers' money. This is oft repeated by modern politicians. See https://www.vogue.co.uk/gallery/margaret-thatcher-most-famous-quotes.
5. It's routine to hear modern politicians refer to hard working taxpayers' money.
6. Kuenssberg, L. 2018. No magic money tree. BBC News. See https://www.bbc.com/news/uk-politics-44524605
7. Figure 4.1 is a simplified version of the UK National Balance Sheet. Available at https://www.ons.gov.uk/economy/nationalaccounts/uksectoraccounts/datasets/thenationalbalancesheetestimates. Note that the figures are rounded.
8. Johansson, A., C. Heady, J.M. Arnold, B. Brys and L. Vartia. 2008. Taxation and Economic Growth. OECD Economics Department Working Papers. No. 620, 2008. https://www.oecd.org/tax/tax-policy/41000592.pdf
9. See OBR (UK Office for Budget Responsibility). A brief guide to the public finances at https://obr.uk/forecasts-in-depth/brief-guides-and-explainers/public-finances/.
10. Li, D.X. 2000. On default correlation: a copula function approach. The Risk Metrics Group Working Paper No. 99–07 http://www.cyrusfarivar.com/docs/li.defaultcorrelation.pdf#:~:text=David%20X.%20Li%20April%202000%20Abstract%20This%20paper,as%20the%20correlation%20coef%EF%AC%81cient%20between%20their%20survival%20times See also End Note 10 of Chapter 4.
11. See GDP data for the UK at https://www.ons.gov.uk/economy/grossdomesticproductgdp
12. A random variable has values assigned by the outcome of an experiment. They are designated by letters and may be discrete (having specific values), or continuous (having values along a continuous range).
13. A bell shaped or normal distribution is one with a distribution that looks like a bell sitting on top of a set of numbers. The peak of the curve is the mean, median and mode of the distribution, and the width—measured by a measure called the standard deviation captures the spread of the distribution so that roughly 95% of all cases occur between –2 and +2 standard deviations from the mean. The height of the bell indicates the chance of a number being selected at random. So, for example, let's imagine that we collect the heights of 100 people chosen truly at random. Now we draw a graph of the numbers of people in each height interval of say 5 cm. Let's assume that the smallest person is 170 cm and the tallest is 190 cm. The graph you have drawn is now a block chart of numbers of people in that height interval. If we then fit a best curve to those numbers of people in each height interval, we will get a bell curve.

14. A random sample is one where the members of the sample are chosen with an equal chance of selection. Chance is modelled by mathematicians using the theory of probability.
15. A standard deviation is the positive square root of the mean of the square of deviations of a set of observations from their mean.
16. Group think happens when individuals in a group, team or organisation tend to accept a way of thinking that is a group consensus whether it has been thoroughly tested and exposed to criticism.
17. For an introduction to complexity theory and resilience. See http://www.complexity.soton.ac.uk/theory/_Resilience_Theory.php#:~:text=Complexity%20theories%20have%20strongly%20influenced,study%20of%20real%20world%20systems.
18. When collecting sample data how do we allow for the possibility of unknown unknowns—things you don't know that you don't know as quoted by Donald Rumsfeld at https://www.brainyquote.com/quotes/donald_rumsfeld_148142
19. How the brain controls heartbeats, our digestion and other sub-conscious functions is beyond the scope of this book. The brain somehow matches all of our brain patterns such as the shape of a dog and the sounds of barking and form the concept of the 'dogginess' of a dog with many different breeds and attributes such as noses, ears, tails, etc.
20. The dependability of the answer we get depends on four sufficient conditions on this process. See Blockley, D.I. 1980. The Nature of Structural Design and Safety. Ellis Horwood, Chichester. Available at https://www.scribd.com/document/447204299/Structural-Design-Safety-By-Professor-David-Blockley-pdf
21. See Edward de Bono in Chapter 1 Section Life is both simple and complex.
22. For UK government statistics see https://www.ons.gov.uk/economy/governmentpublicsectorandtaxes/publicsectorfinance/datasets/financialstatisticsforpublicsector
23. The Board use these data to make decisions about the company and they are required by law to publish certain data as financial reports. The reports contain company accounts such as an end of year balance, a statement of profit and loss for the year, relevant notes, and a director's report together with an auditor's report.
24. The financial year usually runs from April 1st to March 31st.
25. Coincident indicators change more or less simultaneously with economic conditions indicators.
26: Yield curves are interest rates of bonds of equal credit with different maturities.
27. Consumer durables are manufactured things such as cars and household appliances that have a relatively long life.
28. Net business formations are the processes of incorporating or registration of businesses.
29. The FTSE 100 is an index based on the share prices of the largest 100 companies.
30. The Dow Jones Industrial Average is an index based on 30 prominent US companies.
31. The S & P 500 is an index of 500 large US companies.
32. The UK house price index is published by the ONS (Office for National Statistics) every month. See https://www.gov.uk/government/collections/uk-house-price-index-reports
33. The Industrial Production Index (IPI) is published monthly. It shows the growth of various sectors in an economy such as mineral mining, electricity, and manufacturing. See https://tradingeconomics.com/united-kingdom/industrial-production

34. The rate of employment/unemployment published is monthly in the UK by the ONS (Office for National Statistics). See https://www.ons.gov.uk/employmentandlabourmarket/peopleinwork/employmentandemployeetypes/timeseries/lf24/lms
35. Other bilateral mathematical relations are equals, greater than, congruent and orthogonal.
36. Linear regression is a statistical technique for fitting a straight-line relationship to two possible quite scattered variables. Some people liken scattered variables to a '*plum pudding*' graph and the linear fit may be misleading.
37. For practical wisdom see Blockley, D.I. 2022. Creativity, Problem Solving and Aesthetics in Engineering, Springer.
38. Arbitrage is trading to take advantage of differing prices for the same asset, e.g., gold on different stock exchanges or solvency. See https://www.investopedia.com/terms/a/arbitrage.asp#:~:text=Arbitrage%20is%20the%20simultaneous%20purchase,markets%20or%20in%20different%20forms.
39. Working capital ratio is current assets/liabilities. It indicates an ability to pay debts.
40. Quick ratio is current assets minus inventories/liabilities. It indicates a liquid covering of liabilities.
41. Earnings per share (EPS) is the net income earned per share of company stock.
42. Price-earnings (P/E) is the share price/EPS.
43. Debt-to-equity is the debt/financial values of shares. It checks if borrowing too high.
44. Return on equity (ROE) is net earnings after tax minus preferred dividends that must be paid after equity. It indicates profitability.
45. The supply of money is published monthly in the UK by the Bank of England. See https://www.bankofengland.co.uk/statistics
46. For UK business investment data see https://www.ons.gov.uk/economy/grossdomesticproductgdp/bulletins/businessinvestment/previousReleases
47. For UK data on public spending see https://www.gov.uk/government/statistics/public-spending-statistics-release-july-2022/public-spending-statistics-july-2022 can cause a rise in overall demand for goods and services but that can also lead to *inflation.*
48. Crowding out, or pushing aside, occurs when increased government spending drives down spending in the private sector.
49. The output multiplier is the ratio of investment or spending to economic output.
50. Treasury, H.M. 2002. Long-term public finance report: an analysis of fiscal sustainability. See http://fmwww.bc.edu/repec/mmfc03/Tam.pdf

Chapter 5

Sorting the Wheat from the Chaff

Anyone who has never made a mistake has never tried anything new
—Albert Einstein

***Long ago a Woodman was feeling very happy*[1]**. *He was about to marry his childhood sweetheart and live happily ever after. One day whilst working in the woods he came across a Wizard. The Wizard said, "I am your guardian and I have come to make you an offer that will make your life much fulfilled". "What is it?" asked the Woodman quizzically. "The offer is this" was the reply. "You can choose to marry a beautiful princess who is very rich, or you can marry your sweetheart. I promise that if you marry the princess, you will become a prince and live a life of luxury". "What's the catch?" asked the Woodman. "There isn't one said the Wizard. But you will not meet the princess until she is unveiled after the wedding". The Woodman had always been poor. So, despite his love for his sweetheart, the life of a prince was so attractive that he accepted the Wizard's offer. It came to pass that the princess was indeed beautiful, but she was very bad tempered and always grumbling. If anything wasn't to her liking, she blamed her new husband. The Woodman, now a prince, was very rich but became very unhappy. His sweetheart was heartbroken but after a while married the village blacksmith and they lived happily ever after.*

It is better to live happily in poverty than be rich in misery.

The Hedgehog and the Fox[2]

A fox crossed a river but unfortunately got its tail entangled in a bush. He couldn't move. Mosquitoes settled on the tail and enjoyed a good meal. A hedgehog took pity on the fox. "You are in a bad way, Master Fox", said the hedgehog; "do you want me to relieve you by driving off those blood sucking mosquitoes?"

"Thank you, Master Hedgehog", said the Fox, "but I would rather not".

"Why not"? asked the hedgehog.

"Well, you see", was the answer, "these mosquitoes have had their fill; if you drive them away, others will come with fresh appetite and bleed me to death". Better to bear a lesser evil than to risk a greater in removing it.

The Museum Guide

A guide was describing an old relic to a group of tourists. He said, "This relic is 10,003 years old". "How do you know that so precisely?" asked one tourist. The guide replied, "because an expert came and did some tests and pronounced that it was 10,000 years old—and that was 3 years ago".

SORTING THE WHEAT FROM THE CHAFF

When we were hunter-gathers around 6,000 years ago an individual human might have encountered around a thousand people during a lifetime. Now in a big city like New York you might pass that many people in half an hour. Some scientists estimate that we are bombarded every day with more data than people were exposed to in a whole lifetime only 100 years ago. Life seems to be getting faster and more frenzied by the day. Media devices such as TV, radio, tablets, cell phones, and computers are now our constant companions, and they get updated with more 'bells and whistles'. We get to expect to experience more things at a breakneck pace. We get annoyed when the screen freezes or nothing moves. And we are aware that data about us is being compiled as we browse. Most of us have no knowledge of what information is collected, how it is used and the impact of various algorithms, clouds, networks are having on our individual and social lives.

The mystery is almost a new kind of omniscience that, like God's secrets, is unknowable to ordinary human beings. And as these technologies reach out into every part of our lives their mystery will be almost impossible to regulate. We are living in an era of a mythology of 'big data'. There is a widespread view that large data sets offer a higher form of intelligence and knowledge of unprecedented truth, objectivity, and accuracy. Insights can be generated that were previously impossible. Like all myths, there is some truth in this, but we must be very careful and wary of the pitfalls with the likes of so-called fake news.

These pitfalls are centred around the question, 'How do we know who and what to believe and what to reject?' How do we separate the truth from fake news? How do we sort the wheat from the chaff? These are some of the biggest problems of the digital information revolution we are experiencing in the early part of the 21st century. And it is important because in a modern democracy we decide to vote on the information we hear. We must decide who to vote for based on our understanding of very many interacting complex issues—whether about the politicians themselves, their lives, their policies, what we think they can do for us individually and collectively.

Human history has always been plagued by tribalism, propaganda and misinformation says Professor Josh Introne of Syracuse University, School of Information Studies[3]. Belief systems are cultural objects—they occur within a sub-population that shares a cultural identity. Social media makes it easier to dwell in our social networks with the 'best' stories that amass large numbers of followers. And the best stories aren't necessarily true—they may be insular, alarmist, divisive and simply wrong. Through these 'best' stories social groups build separate realities which can become sacred and unassailable. Some unscrupulous politicians exploit our prejudices to get elected, sometimes with messages of animosity and hatred towards minorities and marginal groups.

So, what do we do about it? First, we need to educate people to reject rational ignorance that causes harm either to individuals or groups. Animosity and hatred have no place in principled capitalism and democracy. Second, we should recall from the last chapter, why we discussed measurement, evidence, statistics, and complexity. Remember that science is testable knowledge. The best way to test something is to measure it. Under the right conditions the results are dependable, clear, and repeatable. Measurement is what makes science so important. But, as we have seen earlier, in the social sciences measurements are not so dependable. Data is statistical and difficult to interpret. Before we use any information to act it is crucial to check that data is interpreted within the context in which it was gathered. Much of the data we are bombarded with day to day is not scientific and probably not well tested.

So, in day-to-day discourse, if the messages we are given rely on models that are beyond our understanding, then we must satisfy ourselves that the messenger is qualified and reliable. We should look to see if the information has been tested by critical discussion between many different people. If we do that then gossip and ill-founded comment is immediately ruled out. We have to know that good quality media is impartial. That means we should expect and support only high-quality journalism. We should refuse to buy newspapers or watch any other media that is not committed to reporting the news impartially.

In sum we must not just simply accept what we are told. We should always have a good dose of skepticism. If it is important to us (for example, use it to do something) then we must do our best to single out disinformation. Some warning signs are fairly obvious. For example, material that is too good to be true, inconsistent, overblown, or dramatic. We should take notice only of sources that we feel can be trusted and we understand their intent. For example, we should only take medical advice from qualified doctors and scientists and not from some crank website. We should only divulge personal data when we are sure it is for a good reason—for example to pay a bill. In other words, we need to evaluate critically the information we get with a level of scrutiny that is appropriate for the context.

We should be tolerant of others who don't share our own beliefs. But we should be prepared to discuss critically the ideas that we hear and think without resorting to personal insults or even violence. All that requires some awareness, understanding, informed knowledge but perhaps most of all an open mind.

High on the list of information most of us have difficulty in understanding is the way the economy works and how it is run by politicians and economists. We tend to leave to the 'experts' because the issues may seem remote from everyday life, we have little direct influence and matters are too complex for us to grasp. Both politics and economics are therefore prone to conspiracy theories and fake news. As we have said, the economy is socially and technically simple and complex at the same time. That can leave it open to simplistic views that are motivated by power, money or other reasons and not thought through. Economics is important to everyone because it is about the kind of society we want and how we go about getting it. It is definitely not remote or irrelevant to our daily lives. Clearly the problem for non-specialists is that we get bombarded with many different views from politicians and pundits who have their own motives. Often, we may not understand these views but what they do say and act upon affects us all. The poorest are frequently the most harmed. This is why some degree of economic literacy is important for each one of us. It is why we need to learn and adjust to modern conditions.

OUR BRAINS ARE BUILT TO LEARN AND ADJUST TO CHANGE

The five senses of seeing, hearing, feeling, tasting, and touching enable us to learn from experience and get better at noticing the things that are important to us and ignoring the less important. This is called 'selective attention' by some scientists. Through it we can (but often do not) learn from our successes and our mistakes. But this becomes difficult when reports of data and events rarely tell us about any assumptions behind the so-called facts. For example, how do we interpret the data that 9.1 million people watched a given TV program? You know you watched it with 3 friends, but no-one asked you or them. Does that mean that the quoted number should be 9.1 million plus four? We are given no indication of the accuracy of the number. It seems as if there is an implied assumption that there is one 'true' answer out there even though common sense tells us otherwise.

Philosophers and scientists give that assumption a name—determinism. Taken literally it assumes that output is determined solely by input and initial conditions. Whilst common sense tells us that viewing figures aren't determined and 'true,' we may have some vague idea that it is an estimate based on a sample. But few of us have little idea how the number was arrived at. TV viewing figures are of interest, but not particularly important, for most of us. But they are crucially and financially important for advertisers, performers, and producers. They know that the quoted viewing figure is not precise—we could treat is as a 'ballpark' figure. But a lack of attention to the uncertainty in data can be misleading and potentially harmful. One economic example is the notion of profit.

HAS THE CAPITALIST SYSTEM GONE TOO FAR IN THE PURSUIT OF PROFIT?

Many people think of profit as a 'dirty word'. Indeed, Russell Brand called it a filthy word[4] It is a classic example of data that can get misinterpreted as meaning something else from what it actually is. In fact, profit is simply the proof that economic value is being created. The real issue is more about who benefits from profit when it is pursued to the detriment of all other values.

Let's consider what profit actually is. We will leave aside, for a moment, two important things. First that value should be more than money and include other forms of capital such as environmental, individual social, or political. Second, we won't consider who benefits from getting that profit.

So, with those provisos, we have the costs of doing something. They may include labour, wages, raw materials, other inputs, and the opportunity costs of doing something else with all of these things. Then we also have the income from having done this thing, whatever it is. Profit is, simply, an acknowledgement that the income is higher than the costs of having done it. Thus, financial value has been created. We very much need profits to be made because they are, by definition, the proof that economic value is being created. Profit is proof that the value of outputs is higher than the value of inputs.

Now we come to the controversy over the interpretation of the word profit that we said derives from two sources. First, who benefits. Second, the pursuit of profit to the detriment of other values. If the directors of the company get greedy, they can pay themselves disproportionate salaries and bonuses if their shareholders allow them to. They can pay dividends to shareholders. But most importantly the business can benefit by retaining the profit—either as cash or by investing into new assets or projects such as coping with climate change.

Too often profit is pursued to the detriment of other values. For example, a company may make people redundant, destroy biodiversity, pollute the atmosphere, or damage whole communities—simply in pursuit of more profit. In the recent past UK utility companies have been privatized. Investment has been limited in order to pay dividends to shareholders. In some cases, for example, river pollution from excess wastewater, this has been to the detriment of the environment. Other services such as the Royal Mail in the UK have been made to pursue profit at the expense of over working its staff and reducing the quality of its postal deliveries. The post is no longer seen as a public service but as a private company that must make a profit.

One of the indicators of the pursuit of profit above all else is the size of shipping. It has had lots of unintended consequences. For example, in March 2022, the ship *Ever Given* ran aground in the Suez Canal for six days. It completely blocked this essential artery of trade. A big boat got stuck and people had to get it unstuck. The ship was massive—twice as long as the Suez Canal is wide at the point where it grounded. It was stacked with 18,000 containers and millions of dollars of goods. The ship was Japanese-owned, German-managed, and Taiwanese-operated, and registered in Panama. The captain and crew were from India. The financial pressure to keep to schedule were so great that the *Ever Given* had only two pilots on the bridge.

The era of container ships began in 1966. Ports were mechanized. Jobs that had once been done by gangs of stevedores to load and unload goods were almost entirely lost. Loading and unloading ships became the pushing of levers in gigantic gantry cranes. Everything became huge. The monstrous infrastructure aimed at making profits with a reduced, deregulated, movement of goods. The oceans became trading arteries. Ships no longer carried shipping manifests—a clerk somewhere else in the world calculates what cargo should be loaded, when and where. Containers can be taken from trucks, lifted by crane, and stacked directly onto ships—standardizing and radically reducing costs. Sailors maintain the lifespan of the ship. The workers are separated from the value of their labour—they have no knowledge of what is in the containers, from microchips to hazardous waste. The container made the world smaller and the world economy bigger.

The enormous size of the *Ever Given* together with the wind speed made steering through the Suez Canal very tricky. The ship veered from one direction to another, the two pilots disagreed and argued about what needed to be done. The ship snaked and became uncontrollable and eventually got stuck. Lloyds estimated that the *Ever-Given* cost world trade US$ 9.6bn a day for the 6 days it took to release it. Clearly the pursuit of profit, no matter the impact on the environment, has limits.

At the heart of all this toing and froing of goods and money in pursuit of profit is spending—consumer spending. Spending and profit go hand in hand.

Without spending companies would have no customers, no sales income and hence no profits. Recall that in Chapter 1, I quoted Jon Alexander[5] who called this the consumer story—summed up as 'Just go shopping'. The consumer story is leading us into a spiral of decline. It assumes we are narrow individuals and not the ones to solve the problems of our world. Our agency is limited to choosing between options someone else provides. All we really need to do is buy better stuff.

As we have already noted the consumer story leads to alternating cycles of boom and bust—economic expansion (growth) and contraction (decline, recession). During expansions—the good times—there is greater production, high returns for investors, more jobs, rising incomes, and hence more spending. During recessions production declines, investors lose money, jobs are scarce, average incomes are falling and spending declines. The length and depth of the cycles vary from several months to several years, with an average length of around 5 years going back to the 1850s. A severe recession can become a depression where the economy has a sustained long-term downturn in economic activity.

Many economists see boom and bust as embedded in the capitalist system. It is easy to see why. If we start during a boom, a central bank has often made it easier to obtain credit by lending money at low interest rates. Individuals and businesses can borrow money easily and cheaply and invest it or spend it and the economy grows. The problem is that when credit is too easy to obtain and interest rates are too low, people over spend and overinvest. Demand will slacken when people have too much debt, so companies reduce production and lay off workers.

The things that have been overinvested then decline in economic value. Investors lose money, consumers reduce spending and companies cut jobs. Credit becomes more difficult to obtain because borrowers can't repay their loans. Confidence drops sharply, people get nervous, and the stock market corrects or even crashes. Investors sell to buy safer investments such as bonds, gold, and the fiat currency. Government subsidies, if ill targeted, can make matters worse by encouraging overinvestment in the subsidized item. For example, the mortgage interest tax deduction subsidizes a home purchase by making the mortgage interest less expensive. The subsidy encourages more people to buy homes which perhaps they can't afford, and this reinforces a downward economic spiral.

We can change to the consumer story into the citizen story of collaboration and cooperation. That would enable us to get involved—to see that we should be asking ourselves and formulating answers to questions such as should we tax carbon emissions? How can we conserve water and energy resources? How can we make or systems more resilient?

THE CITIZEN STORY

Jon Alexander's citizen story of collaboration and cooperation may seem idealistic and impractical. But he has given many examples where, around the world, the narrative is changing. For example, Alexander highlights Kennedy Odede, a man who started with a football and street theatre in one of the slums of Nairobi. His organisation called *Shining Hope for Communities* has grown so much that it has enabled over two million slum dwellers to support one another through the pandemic. It even plays host to a nascent World Communities Forum, a collective alternative to the World Economic Forum in Davos.

If we take on the citizen story, then we see immediately that all of us working together are smarter than any of us working alone. We see that finding our way through the complex challenges of the 21st century requires us to embrace the ideas, energy, and resources of us all.

Alexander's form of citizenship goes far beyond national boundaries. It is the deeper meaning of citizen as *together people*—recognising our interdependence and need for community. It's an active verb rather than a noun—a notion of doing something. Citizens look around, identify the domains they can influence, find collaborators, and engage with them. And we need our organisations to allow and encourage us to do so. The citizen story is a bigger story of who we are as human beings. Notable examples abound—volunteering is especially important. The story of the volunteers who in June 2018 rescued 12 boys from caves in Thailand's Chiang Rai province together with their football coach was a remarkable example of friendship, human endurance, and the lengths some people will go to save someone else's child[6]. Other examples are the countless volunteers who helped vulnerable people during the Covid-19 pandemic[7] and those unsung heroes who save the lives of others by donating their healthy organs. The valour shown by those who intervene when they witness a violent attack regardless of the danger to themselves. The heroes of the French Resistance in World War 2, the bravery of Ukrainians against the Russian invasion, and the 260 villagers of Eyam in Derbyshire England who in 1666 consciously didn't leave the village during the bubonic plague and sacrificed themselves to save the lives of those in neighbouring towns and villages[8]. The list is endless yet historically ignored in economic thought.

Volunteering is any unpaid work that someone does for the benefit of others or for a specific cause. Volunteers want to give something back to society and help others. It is personally satisfying and rewarding especially when others appreciate what you do. It creates new networks and relationships and serves the interests of the community. The Covid-19 pandemic has changed behaviors. Daniel King, professor of organisational behavior at Nottingham Business School says[9], '*The pandemic has seen a shift in not only who has been volunteering but also how they are volunteering*'. Alex Farrow, head of networks and influencing at the UK National Council for Voluntary Organisations, says: '*Overall, we see a mixed picture for formal volunteering during the pandemic......The increase in volunteer diversity during the pandemic is welcome and overdue*'.

Many businesses now talk in terms of stakeholder value not shareholder value. Charities and non-profit organisations encourage participation and a more collective approach. A platform called Restor[10] allows grassroots nature conservation projects from all over the world to plot their impact, connect and collaborate. But for these ideas to spread we must, as Alexander says, neither accept what we are given as the only possibility, nor throw our toys from the pram when we do not like what is on offer. Citizens propose—they do not simply reject. Citizens accept responsibility, and create opportunities for each other—they step up, and step in. They believe that to change something, you must create new models that make existing models obsolete. They help each other to avoid disinformation and fake news.

Leadership is crucial to the citizen story. If those in positions of power act as if there is nothing wrong, we mistrust them even more. Citizen leaders begin by acknowledging uncertainty—we are sadly lacking them in politics. They share questions and challenges *with us* rather than most current politicians who provide answers *for us* which often fail. Citizen leaders create opportunities for us to participate, contribute and develop the learning power I referred to in the Prologue. They acknowledge and don't deny unknowns and more importantly unknown unknowns. They don't pretend to know what the future will look like, but they reassure us that we should build it by working and learning together.

Some scientists estimate that we are bombarded every day with more data than people were exposed to in a whole lifetime only 100 years ago. Life seems to be getting faster and more frenzied by the day. How do we sort the wheat from the chaff? First, we must not accept what we are told without examining it to single out disinformation. We should avoid material that is too good to be true, inconsistent, overblown, or dramatic. We need to look at only trusted sources and try to understand their intent—for example, taking medical advice from qualified doctors and scientists and not from some crank website. In other words, we need to evaluate critically the information we get with a level of scrutiny that is appropriate for the context. We should be tolerant of others who don't share our own beliefs and be always prepared to discuss critically the ideas that we hear and think. All that requires some awareness, understanding, informed knowledge but perhaps most of all an open mind.

Our brains are built to learn and adjust to change. The five senses of seeing, hearing, feeling, tasting, and touching enable us to learn from experience and get better at noticing the things that are important to us and ignoring the less important. We can (but often do not) learn from our successes and our mistakes. But this becomes difficult when reports of data and events rarely tell us about any assumptions behind the so-called facts.

Has the capitalist system gone too far in the pursuit of profit? Many people think of profit as a 'dirty word'. In fact, profit is a requirement for a successful company. It is simply the proof that economic value is being created. The issue is more about who benefits from profit when it is pursued to the detriment of all other values. Value should be more than economic exchange of worth as money but needs to include other forms of capital such as environmental, individual social, or political. One of the most quoted measures of the financial value of the production of goods and services in a country during a certain period is GDP or Gross Domestic Product. And yet it is widely recognized as inadequate, as it includes costs and waste, i.e., activities that may be unproductive or even destructive. Economists rely on a wide use of ratios to interpret financial data. It is all too easy to underestimate the contextual nature of such data and the difficulty in interpreting it. Judgement and experience are required.

The idea of citizenship goes far beyond national boundaries. It is the deeper meaning of citizen as *together people*—recognising our interdependence and need for community. It's an active verb rather than a noun—a notion of doing something. Citizens look around, identify the domains they can influence, find collaborators, and engage with them. And we need our organisations to allow and encourage us to do so. The citizen story is a bigger story of who we are as human beings. It is a story of proactivity, of being creative, ingenious, positive, and learning together. So, in Part 3 we turn to look at how we act to build trust, mange uncertainty and create better outcomes for us all.

End Notes

1. Aesop Fable: Long ago a Woodman was feeling very happy.
2. Aesop Fable: The hedgehog and the fox.
3. Syracuse University, School of Information Studies. 2021. How our Belief Systems Make Us More Susceptible to Misinformation. See https://ischool.syr.edu/how-our-belief-systems-make-us-more-susceptible-to-misinformation/
4. Worstall, T. 2013. On Russell Brand's misunderstanding of profit. Adam Smith Institute. See https://www.adamsmith.org/blog/economics/on-russell-brand-s-misunderstanding-of-profit/

5. See Chapter 1 John Alexander's Citizen's Story. Alexander, J. (2022). Citizens: Why the Key to Fixing Everything is All of Us. Canbury, UK.
6. Cheung, H. and T. Wong. 2018. The full story of Thailand's extraordinary cave rescue. BBC News. https://www.bbc.com/news/world-asia-44791998
7. British Red Cross. 2020. Community reserve volunteers helping in Brixton. See https://reserves.redcross.org.uk/2020/07/03/community-reserve-volunteers-helping-brixton/
8. McKenna, D. 2016. Eyam plague: The village of the damned. BBC News. See https://www.bbc.com/news/uk-england-35064071
9. Nottingham Trent University. 2021. Research reveals the rise of the digital volunteer, with more than 90% of voluntary organisations moving operations online during the pandemic. See https://www.ntu.ac.uk/about-us/news/news-articles/2021/05/latest-research-reveals-mixed-impact-of-pandemic-on-volunteering-numbers-despite-more-positive-outlook,-increased-diversity-and-rise-of-the-digital-volunteer
10. See Restor at https://restor.eco. Transparency and connectivity for restoration and conservation efforts around the world. A global network of actors is working together to advance critical nature-based solutions. Restor enables the exchange of data, local knowledge, and funds between them.

Part 3

Acting

Preamble Part 3

Excerpts from The Blind Men and the Elephant[1] *by John Godfrey Sachs.*

'It was six men of Indostan, to learning much inclined, who went to see the Elephant (Though all of them were blind), that each by observation, might satisfy his mind...

One man touching the body, thinks it's a wall. Another feeling the tusks, thinks it's a spear. The man touching the trunk, thinks it's a snake. Another feeling the leg, thinks it a tree trunk.

The man touching the ear, thinks it is a fan. Another feeling the tail, thinks it's a rope.

...And so, these men of Indostan, disputed loud and long, each in his own opinion, exceeding stiff and strong. Though each was partly in the right, and all were in the wrong...'

In Part 1 we saw how we become what we ***think***. In Part 2 we learned how to ***test*** what we are told. Now in Part 3 we must focus on how we ***act***. We must act to build trust. We have to 'get to grips' with uncertainty and risk. We should find ways to do what Karl Popper called 'piecemeal social engineering'. Feeling our way towards a better future by reform and not revolution. Popper wrote[2] the piecemeal social engineer tries to achieve his ends by making '*small adjustments and readjustments which can be continually improved upon...Accordingly, he will make his way, step by step, carefully comparing the results achieved, and always on the lookout for the unavoidable unwanted consequences of any reform; and he will avoid undertaking reforms of a complexity and scope which make it impossible for him to disentangle causes and effects, and to know what he is really doing'.* I will propose that to do this we should be nurturing what Aristotle called *phronesis* or practical wisdom as mentioned earlier. I think it important to embrace practical wisdom and 'systems thinking'.

Uncertainty and risk are not well understood and are proving big stumbling blocks. We have always wanted to know what the future holds, but we are learning that there are limits to what we can predict. We need to drop the idea of 'predict and provide' and adopt Popper's piecemeal social engineering. We need to nurture a sense of global citizenship that will help reverse and reduce the growing disparity between rich and poor.

End Notes

1. https://allpoetry.com/The-Blind-Man-And-The-Elephant
2. Popper, K. 1957. The Poverty of Historicism. Beacon Press.

Chapter 6

Gimme Money

That's What I Want—The Beatles

Money is Only Important When You Don't Have Any

The cat and the fox[1] *were talking politics. The fox said 'I don't care if things turn out bad, I have a thousand tricks, before I get hurt. But what about you cat? The cat replied, 'I have only one thing I can do and if that doesn't work, I am done for'. 'I am sorry to hear that' says the fox. 'I would gladly help you, but when things get bad it is not good to trust and rely on others—it's everyone for themselves'.*

Scarcely had the fox spoken when a pack of hounds bounded up in full cry. The cat ran up a tree and sat securely on the top branches—he was safe. Despite his thousand tricks the fox could not escape. The dogs tore him into pieces.

A little common sense is often of more value than a lot of cunning.

An Old Man and his Gold

An old man sells all his goods for a large lump of gold. Then he buries the gold in a hole just near his house. Every day he uncovers the gold, admires it for a while and then covers it back up. A robber notices this ritual. He follows the old man, sees the buried gold, then when the old man has gone back home, digs up the gold and steals it. The next day the old man finds that his gold his missing and cries out in agony. A neighbour hears the old man's cry and sympathizes. He tells the old man to put a rock in the hole and cover it back up—saying "It makes no difference, does it? You didn't do anything with the gold anyway".

Man: *"I have decided to boycott shopping at AB plc".*

Woman: *"Why's that?"*

Man: *"Because they only sell things I can't afford".*

MAKE YOUR MONEY WORK FOR YOU

Banks and financial advisors constantly say, 'make your money work for you'. Essentially use your money to make more money. More broadly take control of your finances to improve the stability and security of your income and expenditure. If you have enough money (many people do not) then invest it to build your wealth and be financially independent. Of course, to do any of that you must first understand where your money comes from and where it is going. Then you can find better ways to use it. Whether you have enough money to invest or not the usual advice is to learn to budget, remove bad financial habits, pay off and avoid debt, prioritize spending, save for the future, have an emergency fund, and invest for your future.

If you store your money under your mattress, it loses economic value. As time goes by, you will be able to buy fewer goods with it. Your purchasing power will reduce. Although you have the same amount of currency you will effectively have a smaller amount of money to spend. Recall that we said that spending is at the root of our current form of capitalism. Money makes more money and keeps or increases its economic value. But you have to invest it by buying things that appreciate in economic value, ranging from artwork to government bonds. Investing makes you wealthier and investing depends on interest rates. So, you should always be aware of the current interest rates.

MONEY AND WEALTH ARE DIFFERENT

As we noted in Chapter 1 money and wealth are different. Money comes in many forms that economists call M0 through to M4 depending on its liquidity or ease of turning into cash. Wages and salaries used to be paid in cash, often weekly. Now it's usually paid electronically direct into your bank account. Money gives you purchasing power—it is an instrument of exchange. The more money you have the wealthier you feel. But wealth is not defined that way. Economists say that wealth is the excess of your assets over your liabilities. The relationship between wealth and money is another thing which is both simple and complex at the same time.

It is simple in the common sense that the more money you have the wealthier you become if you invest it to maintain its purchasing power. But it is complex as the story of a castaway on a desert island in Chapter 2 shows. She finds a gold mine and a buried case full of bank notes. But the finds do not make her wealthy. What the story demonstrates is that economic value depends on exchange value or utility. That is why economists have developed and used theories of utility to guide their thinking.

INVESTING

If you, a company, or government invest in a bank savings account, a new building or machinery, or infrastructure, then you forgo spending that money on something else for the moment. You expect to spend it sometime in the future. You will recall from Chapter 2 that investments and interest rates have a checkered history. Thomas Aquinas condemned the moneylenders as usurers who would go to hell with thieves and robbers. But then over time the church began to accept the need for people to earn interest on their investments.

Interest rates are crucial to a capitalist economy because they are the factor that makes your money work for you. Wealth derives not from wages but from investments. In a sense interest is the economic time value of money—the price of time[2]. That is because the returns on your investments will take time to come about. Interest represents the difference between the economic value of a sum of currency at a point in the future compared to its current value. In theory the rate of return should be at least equal to the investors time preference—the length of time people will wait for their return. The economist Friedrich Hayek (1889–1992) said that when interest rates decline businesses tend to invest in projects with more distant payoffs. Many economists think that there is a natural, neutral or equilibrium rate of interest where individuals lend and borrow freely. That natural rate reflects the time preferences of society so that we neither borrow too much or save too little and ensures that capital is used efficiently. It's a rate that is not so low that it subsidizes bankers. Nor is it too high as to hurt borrowers. It is difficult (if not impossible) to decide what the natural rate is, but we can tell when we are not at it[3]. Hayek maintained that if interest rates are kept below their natural level, then there is a tendency towards malinvestment where outlays are not justified. Consumers are also affected by low interest rates. Cheap credit encourages households to take on too much debt. They borrow and spend, bringing forward their consumption forward, but when the time comes to pay, they can get into trouble. Hayek thought that widespread malinvestments inevitably end up in an economic crisis.

He would not have been surprised therefore by the events leading up to the financial crisis of 2008. The build-up of excess amounts of consumer debt were disastrous. The unintended consequences of ultra-low interest rates in the twenty-first century were inflated house price bubbles around the world[4]. The demand therefore lowered as households pulled back on spending in order to pay back credit. Many industries had excess capacity. After 2008 corporate debt soared. But the loans were used for financial purposes including share buybacks and company buyouts. It was a takeover boom that reduced competition and fostered monopolies. Consequently, investment lowered as did economic growth.

The main beneficiaries of the great wealth bubble were the richest 1 per cent. The rest of us struggled with weak growth of income, unaffordable housing, and excessive debt. That exacerbated the already rising inequality—not good in any sense for anyone. Inequality is not good for economic growth. It is especially not good when it arises from corporate rent seeking of the type which thrived after 2008. Economists use the term rent to describe economic wealth obtained through shrewd or potentially manipulative use of resources. For example, when a company lobbies the government for grants, subsidies, or protection from tariffs. Rent seeking is increasing one's share of existing wealth without improving productivity. Rent-seeking results in reduced economic efficiency through a poor allocation of resources, reduced actual creation of wealth, lost tax revenue, and increased income inequality.

When interest rates are too high and the cost of borrowing is set at too high a level, then people cut back on spending and company profits fall. Businesses don't borrow to invest, creditors (the lenders) gain unduly at the expense of debtors (the borrowers). Financial values of capital assets fall, workers remain idle, and the economy stagnates. When the yield from bonds exceeds the growth in national income, then existing debt becomes a real burden. Bankruptcy is then on the horizon[5].

FISCAL AND MONETARY POLICIES ARE BLUNT INSTRUMENTS–WE NEED BETTER AUTOMATIC STABILIZERS

A blunt instrument is a metaphor for something that is not sufficiently precise to be effective for its intended purpose. As we have seen in earlier chapters governments set fiscal policy with only three main instruments, tax, borrowing and spending. However, we must recognize that there are already automatic stabilizers that temper the economy when it overheats and stimulates it when is slumps—without policy intervention. For example, as Stephanie Kelton[6] points out '*in a slowing economy, more people become eligible for unemployment insurance and food stamps, so government spending automatically increases in those categories of the federal budget. Meanwhile, tax receipts automatically decline as the weak economy leaves corporations and individuals with less taxable income. With higher spending and lower tax revenue, the federal deficit automatically increases. The bigger deficit cushions the economic slowdown. In a booming economy, the reverse happens*'.

The three biggest US packages ($2.2 trillion in March 2020, $900 billion in December 2020, and $1.9 trillion in March 2021) were all enacted in middle of a global pandemic. The US Inflation Reduction Act of 2022 contains US$ 500 billion in new spending and tax breaks that aim to boost clean energy, reduce healthcare costs, and increase tax revenues. Policy makers erred on the side of doing too much as opposed to doing too little. Kelton urges governments '*to avoid cobbling together multi-trillion-dollar fiscal rescue packages in a state of panic*'. One way is to strengthen automatic stabilizers such as a jobs guarantee scheme as mentioned in Chapter 2[7].

Central banks really have only two instruments to influence monetary policy, interest rates and the setting of minimum bank reserves. The reason economic forecasting is so difficult and uncertain is because the economy depends on the collective intentions of millions of people. This in turn depends on their confidence in that economy and their behavior, which as we saw in Chapter 3 is a complex mixture of the rational and irrational.

Nevertheless, fiscal, and monetary policies are important, and the blunt instruments do have a role. In this chapter we are going to look at monetary policies and in particular interest rates, in some depth.

BANKS CAN CHANGE THE AMOUNT OF MONEY IN CIRCULATION

Economists refer to the amount of money in circulation as the money supply[8] Changes of the money by banks can be more immediately effective than the more slow-moving changes in fiscal policies.

Bank deposits are most of the money held by the public. But where this stock of bank deposits comes from is often misunderstood. If you go to a bank and ask for a loan, they check you out and hopefully agree. That creates new money. It's an exchange of promises—you promise to pay back the loan, the bank credits your account and allows you to spend the money as you want. Banks actually create deposits rather than lending out savings. In other words, they are not just intermediaries which lend out the deposits that savers have deposited. In practice saving in a bank does not increase the funds available for banks to lend. This is not how many typical textbooks describe what happens. Commercial banks create money, as bank deposits when they make a new loan. We just have to look at the balance sheets of the borrower (individual or company), the bank and the central bank to see what happens. New deposits increase the assets of the consumer (households and companies), and new loans increase their liabilities. New broad money is created. Both sides of the bank's balance sheet increase. For the bank, the created new money equals the amount of new lending. (Although in practice they may be slightly different due to other small factors). So even though new broad money has been created on the consumer's balance sheet there is no change in the amount of 'base money' at the central bank. Base money is the currency issued by central banks as notes and coins. It is held in the reserve account of the commercial bank at the central bank. The larger the stock of deposits then banks may be required, to hold more money at the central bank. This is because more withdrawals may be made by the public or there may be a need to make more payments to other banks. Reserves are, in normal times, supplied 'on demand' by the Bank of England to commercial banks in exchange for other assets on their balance sheets. The aggregate quantity of reserves does not directly constrain the level of lending or creation of deposits by commercial banks.

In sum, every time you deposit a £ or $ into a bank account, the total reserves of that bank increases. The bank then is required to keep some of that money in its reserves at the central bank. But it will lend the rest out to others. As the loan is made, the money supply increases. Recall from Chapter 1 that money comes in several forms from M0–M4. When economists refer to the money supply, they mean M0 (UK) and M1 (USA) the currency held by the public and checkable deposit balances in banks.

The main tools of a central bank are open market operations and quantitative easing or QE. The latter is used less often—and usually only at times of financial crisis. The opposite to QE is QT or quantitative tightening. Open market operations are the buying and selling of securities by a central bank on the open market. Quantitative easing occurs when a central bank creates money electronically out of thin air—then uses it to buy assets. Typically, these are government bonds bought from the market when interest rates are low. Both tools increase the money supply. This encourages other banks to lend more. Consequently, the price of bonds rises, and interest rates are lowered. That means businesses and individuals will borrow, invest, and spend more and so boost the economy.

Central banks use open market operations to influence the supply of money by the buying and selling of government securities (bonds or IOUs). They buy using newly created money deposited in a bank's account with a central bank. Similarly, the central bank can sell securities and take money out of circulation, putting upward pressure on interest rates. Open market operations also include moving interest rates up or down, lending cash to banks, and changing the level of reserves required by banks. If, during a crisis, open market operations are insufficient (e.g., when interest rates are already extremely low) central banks might use quantitative easing to buy a massive variety of securities at large scale. The first central bank to do this in modern times was the Bank of Japan in 2002. USA, UK, and the Eurozone followed suit after the 2008 banking disaster.

Government Bonds

In the UK Government bonds are called gilts or gilt-edged securities and they are secure. In the USA they are called Treasuries. There are two types of securities—conventional and index-linked gilts. Conventional gilts are the simplest and are around 75% of the total. They are a guaranteed liability of the Government to pay the holder of the gilt a fixed cash payment (called a coupon) every six months until the maturity date. Then the holder receives the final payment and the amount purchased is returned. Index-linked gilts differ from conventional gilts in that the semi-annual coupon payments and the principal are adjusted in line with (in the UK) Retail Prices Index (RPI)—to take account of inflation. The UK and US governments have never failed to make payments when they are due.

A government issues bonds in an auction conducted by the central bank with bids from primary dealers. Primary dealers are banks or other financial institution that have been approved to trade securities with a national government. They must prove that they are safe by being able to meet specific requirements for liquidity and quality. In many countries, primary dealers are the only financial entities who can bid for newly issued government securities. They then sell the bonds on at a slight markup, often to pension funds, other banks, and individuals in a bond market. Interest rates affect this market and the demand for bonds. If interest rates are lower than the rate of a bond, demand for the bond will rise because it is a better investment. But if interest rates are above the rate of the bond, demand will drop. Newly issued government bonds are priced according to current interest rates. High inflation means that a fixed payment becomes less valuable over time. The central bank will often increase interest rates thus lowering the market price for a bond because interest rates and bond prices are inversely related. Bonds that pay out in a foreign currency may also drop in value if the foreign exchange rate falls.

A central bank can also increase the money supply by lowering the reserve they require banks to hold. This allows the banks to lend more money. Raising the banks' reserve requirements decreases the supply of money. As mentioned earlier the central banks can conduct open market operations to buy and sell government bonds. Buying bonds increases the money supply and lowers interest rates, which in turn, generates more investment and puts more money in the hands of consumers, and stimulates spending. Businesses order more materials and increase production. This then raises the demand for labour. If the money supply is reduced the opposite occurs.

Keynes thought that government interventions are needed to avoid recessions. He wanted savings to be invested because that would create economic activity and growth. In other words, government spending is a means of stabilizing the economy rather than controlling the money supply. Unfortunately, before the financial crisis of 2008 many economists thought that government should reduce company taxes and loosen restrictions to encourage the employment of more workers.

BOOM AND BUST

Gordon Brown's famous mantra during his 10 years as UK Chancellor of the Exchequer was there would never be a return to boom-and-bust economics. Such was the overconfidence of many in the world of economics. As John Lanchester writes[9] '*It's been said that the four most expensive words in the world are—this time it's different*'. Economic cycles seem to be inevitable—expansions followed by contractions or recessions. History tells us they are recurrent but not regularly periodic and therefore hard to predict. They arise from changes in production and trade. Boom periods are often also called economic bubbles. They occur when prices of goods soar far above their intrinsic value or real worth. Then there follows a bust as prices fall. The so-called business cycle is an example of an economic cycle with four stages of varying lengths of time and severity. First economic activity slows down—a time of contraction. Second, the economy reaches a trough and signs of recovery begin to show. Third, economic activity starts to pick up—a time of expansion. Fourth, expansion reaches a high peak, and economy begins to contract again. As I said earlier, this cycle seems to be a key characteristic of capitalist economies. During the expansionary boom the economy grows, jobs are plentiful, and the market brings high returns to investors. Central banks lend money at low interest rates. Borrowing is easy and cheap and can be invested in, for example, technology stocks or houses. During this time, many people earn high returns on their investments, and the economy grows. But there are risks. When credit is too easy and interest rates too low, for too long, people tend to overinvest. Economists call that malinvestment. For example, if too many houses have been built there won't be enough demand and the bust stage will germinate. During the bust overinvestments will lose value, investors lose money, consumers cut spending and companies cut jobs. Credit becomes more difficult to obtain as some borrowers can't repay loans. The bust contracts the economy and it shrinks. People lose their jobs and investors lose money. The bust stages are called recessions and if particularly severe, a depression. Confidence plummets, investors and consumers get nervous as the stock market corrects or even crashes. Investors sell to buy safer investments such as bonds, gold, and the U.S. dollar. Workers, who are consumers, lose their jobs and they reduce their spending. That exacerbates a downward spiral. Eventually the bust stage stops when prices are so low that those investors start buying again. Confidence can be restored more quickly by suitable fiscal and monetary policies.

HEDGING YOUR BETS

One-way investors try to protect their assets is by hedging. Hedging is a way of protecting investments by a kind of insurance—a strategy to manage risk. However, it is not as simple as an insurance policy on your house or car. Investors hedge one investment by making a trade in another. It is a massive form of gambling with huge sums of money. Hedging can reduce potential gains because you pay a premium to buy and sell derivatives, such as options and futures contracts. Options are financial derivatives that give buyers the right, but not the obligation, to buy or sell an underlying asset at an agreed-upon price and date. Derivatives are a form of security that can become very complex and controversial. They are financial instruments whose value depends, as the name implies, upon the value of another asset. There are four main types of contracts for derivatives. First are futures contracts where investors buy and sell at some predetermined prices at some future specified time. Second are forwards contracts which are the same as futures except that the price is agreed at end of the contract. Third are options contracts where the holder has a right but not an obligation to buy or sell. Fourth are swaps or contracts where the parties to the contract agree to exchange their holdings. There are two basic positions in trading derivatives. The first is called a short position where the investor bets that the price of the holdings will go down. The second is called a long position where the bet is that the price will go up.

American-style options can be exercised at any time. European-style options can only be exercised on the end date. There are two types of options named calls and puts. Calls give the buyer the right, but not the obligation, to buy the underlying asset at the agreed price. Put options give the buyer the right, but not the obligation, to sell the underlying asset at the agreed price.

Don't Try This at Home

Investors can speculate by exploiting price changes to sell call options or buy put options. A call option price is called the strike as specified in the contract. Investors buy call options when they believe the price of the asset will increase and sell if they believe it will decrease. The writer or seller of a put option must buy the asset if the put buyer exercises their option. Investors buy puts when they believe the price of the underlying asset will decrease and sell puts if they believe it will increase.

Trading in derivatives is not for the amateur—it is complex and risky. The market in derivatives gets profits not from goods and services but from shares.

In 1973 two mathematical academic economists Fischer Black and Myron Scholes published a paper containing a partial differential equation[10] that would enable investors to calculate the price of derivatives based on the value of the underlying assets. Within a short time, a new era of trading derivatives took off. Traders were using the Black–Scholes equations in a market that in 2020 is measured around hundreds of trillions of dollars. It seems as if the world of trading lost all of its common sense because the trades were far in excess of what the underlying assets were worth. The risks became so great that some individual traders have lost billions of US$. One of the most infamous was Nick Leeson in 1995 when his trading led to eventual losses of £880 million and destroyed Barings, the UKs oldest merchant bank. John Lancaster in his book '*Whoops*'[11] tells of a senior figure in the Treasury and a bank board member who is reported to have said that he had some good news: '*we're no longer going to get involved in things we don't understand'*. The Treasury man added *'We now own his bank*'.

In 1993 a group of mathematical economists (including Myron Scholes) formed a new trading company called Long Term Capital Management (LTCM). Within four years the company quadrupled in value. It had an equity of US$ 4.72bn. But then in 1998 it imploded when Russia defaulted on its foreign debt obligations. It left a US$ 1.25 trillion hole in the global financial system. The reason was that it had an inbuilt risk of that amount through its borrowing, leverage, and various derivatives. What the LTCM directors had missed was the possible effect of unknown unknowns. They confused mathematical probabilities with real world uncertainties and missed the unexpected—the Russian default.

The world-renowned investor Warren Buffett takes a more pragmatic stance. He is reported to not like derivatives because he prefers to know what is going on in the companies in which he is investing. That is common sense of course. British mathematician Ian Stewart said that the Black–Scholes equation was the mathematical justification for the trading and *'one ingredient in a rich stew of financial irresponsibility, political ineptitude, perverse incentives, and lax regulation that contributed to the financial crisis of 2007–08. The equation itself wasn't the real problem,*' but rather its abuse in the financial industry.

Another example of the inappropriate use of models as tools was highlighted by a *flash crash* in 2010 in the electronic market for securities. Computer algorithms were being used to control trading—some of it at exceedingly high frequencies. The algorithms could, in certain circumstances drop prices very rapidly. On this occasion the Dow Jones Industrial Average fell more than 1,000 points in 10 minutes—the biggest drop ever up to that time. Over $1 trillion in equity disappeared. The market did regain 70% by the end of the day's trading so the effects were not as bad as they might have been. The regulators acted quickly. They installed so-called, circuit breakers in the algorithms to try to prevent more flash crashes. However according to some estimates[12], there are approximately 12 mini flash crashes on any given day. Flash crashes are caused by computer algorithms triggering a domino effect. However, they can occur through glitches, and errors, and even by fraud. The chances of flash crashes are much higher in complex interconnected computer networks operating sophisticated algorithms.

Queen Elizabeth II famously queried the professors at the London School of Economics about the 2008 banking collapse, not quite two months after the Lehman Brothers were declared bankrupt. She simply asked, "Why did nobody see it coming?" The professors were caught off guard. Later a group of prominent British economists wrote a letter to the Queen which said that the extent and severity of the crisis had many causes but was principally a failure of the collective imagination of many bright people, both in this country and internationally, to understand the risks to the system as a whole. So now let's look at the problem a little more deeply. It turns out to be all about risk, and the story gets even more complex.

Arbitrage

Arbitrage is about finding ways of removing risk so that whether you bet that the market will go up or go down you will still make a profit. You get your cake and eat it as well. It is the simultaneous purchase and sale of an asset (stocks, commodities, and currencies) in different markets to exploit tiny differences in their prices. It takes advantage of the inevitable inefficiencies in markets. One version of this is an innovative kind of swap called a credit default swap. The following is a summary of John Lanchester's[13] impressively clear explanation of how this works.

Let's call some financial entities whether individuals, companies, or banks by first names Andrew, Ben, Charles, etc., in alphabetical order. Ben needs a loan and Andrew lends him say £100,000 at an interest rate of £1,000 per month to be paid back after one year. However, Andrew knows Ben quite well and isn't too confident that he will be able to repay the loan in a year's time. One day he is talking to Charles, who it turns out, has much greater confidence in Ben. Andrew therefore makes Charles an offer. Charles takes on the risk of Ben defaulting on the loan and will make up the money on the loan in return for a fee of £50 per month. Charles accepts because he bets that Ben won't default and he'll make an easy profit. Andrew is happy because he has effectively insured away the risk of Ben defaulting. That is because if Ben does default Andrew will collect from Charles. In other words, Andrew has got himself a riskless deal. He has his cake as well as eating it. He sees this as a good way of making money. So, he makes more deals. First, he makes a loan to Dan and sells the risk to Eddie. But he doesn't know Dan and Eddie very well still less their financial affairs. But then he reckons that if he makes many more of these deals then if he makes enough of them, he will 'on average' cover his downside fees with his upside profits. In other words, he makes as many deals as he can cover on the expectation that his profits will, on the whole, outweigh his exposure to losses. At the same time unbeknown to Andrew, Charles, and Eddie and all of the others have made similar deals. They have undertaken similar contracts and sold them on to others. The others have then also sold them on. Eventually these many deals permeate a massive part of the whole financial system.

Just before 2007, Andrew was the bank J.P. Morgan, Ben was Exxon, the oil company and Charles was the European Bank of Reconstruction.

But what were the regulators doing? After years of haggling, they accepted the idea that the selling of risk this way was a good thing. They reasoned that it spread the risk out through the financial system rather than concentrating it in one place. The credit default swaps were bundled together in various ways in what came to be known as securitization. As pointed out earlier it relies on safety in numbers and the law of averages. Even if some loans did default the others wouldn't and the stream of revenue would keep going. The firms began to divide bundles into levels of risk with different rates of return—a process called tranching. Special purpose vehicles were set up as offshore shell companies to fulfil the role of Charles.

Then someone had the idea of extending credit default swaps to house buying and mortgages. They also saw a way they could extend mortgages to lower-income groups at higher risk of defaulting. These were unusual 'non-conforming' loans[14] and were known as sub-prime mortgages—a rather unhelpful and misleading name. Because they were higher risk, they attracted a higher rate of interest than conforming prime loans. Securitization promised much to the lenders. Unlike prime mortgages however it was difficult to model the risks statistically. Then David Li, a Chinese working in North America[15] seemed to have solved the problem with some difficult mathematics. He replaced a discrete correlation between variables[16] with a random variable[17] of time-until-default or survival time. He used market rather than historical information to study a multiple credit portfolio. He constructed a mathematical function that mathematicians regard as a simple convenient mathematical approach.[18] He used it to link single variable to multivariable joint probability distributions. He computed pairwise correlations of survival times using previously published correlations between assets over a one-year period and gave some numerical examples. He concluded his paper by stating that there was no unique solution to the problem. His suggestions were quickly turned into a tool for financial institutions to correlate associations between multiple financial securities. Those who did not understand the limitations of the mathematics thought that the tool accurately priced credit default swaps for a wide range of investments. These included mortgages which had been too complex to price. The market went wild. Financiers either did not understand or care about David Li's assumptions and ignored the limitations of the model. The market grew to US$ 62 trillion by 2007. The reason was that they thought, quite against any common sense, that they had found a way of trading where everyone won. Financial Institutions began to put lots of mortgages into a single pool and then divide them up to sell them on to investors at various levels of risk. They thought that the lender, when making a loan to someone needing a mortgage, did not need to worry about the risks of default. That is because he would sell it on. In other words, the lender no longer owned the risk because it had been securitized. But everyone had forgotten one of Murphy's Laws of common sense—that if something can go wrong it will. Not every institution joined in the bonanza. The most obvious exception was J.P. Morgan, the very firm that had been influential in the invention of credit swaps in the first place.

In all the activity an unfortunate change of culture occurred. As John Lanchester[19] puts it, the change was '*from lending to help poor but reliable people to own a house to a manufactured process driven by capital which is set loose looking for people to sign up loans'*. It became a matter for salespeople. It became an epidemic of predatory lending with lenders doing everything they could to sign up borrowers. The already flexible rules were stretched as far as they could be taken. For example, borrowers were asked to declare their earnings and the lender did not check the facts. These are now known as liar loans. Worse still loans were made to people with no income, job, or assets. Lenders didn't care because they knew they were selling on their loans. Loans were being made to people who would never be able to pay them back. The whole market was riddled with scams and tricks. Then in 2004 the 'chickens came home to roost'. Interest rates were raised, and higher mortgage payments squeezed the abilities of borrowers to pay. The bubble finally burst in 2007. The crash in the housing industry drove the U.S. financial industry to its knees. The USA banking industry is global in its effects and almost pushed the world's financial systems to near collapse. Governments around the world were forced to implement enormous bail-out programs for financial institutions judged to be too big to fail. The investment bank, Lehman Brothers, collapsed in September 2008. A toxic mixture of deflation, high unemployment, and soaring government debt had been created. Interest rates were pushed down, and governments acquired large national debt through quantitative easing as they sought to rescue banks. Unfortunately, the newly printed money inflated financial assets rather than consumer prices. In Europe GDP per capita declined in a way not foreseen by central bankers[20]. The very low interest rates discouraged investment and savings. In 2016, Mervyn King, former Governor of the Bank of England[21] referred to a paradox of policy where interest rates were too high to permit rapid growth of demand in the short run, but too low to be consistent with a proper balance between spending and saving in the long run. Chancellor[22] comments '*In the short run, ultra-low interest rates boosted consumption by substituting bubble wealth for savings, but in the long run it was a disaster. Savings are needed for the accumulation of capital. Societies that don't invest enough are doomed to stagnate*'. A Bank of England study[23] found that the decline in interest rates was mostly responsible for ballooning pension deficits. Defined benefit pension plans with their copper-bottomed promise of a given retirement income were closed to new members and replaced by defined contribution pension schemes which gave no guaranteed income on retirement.

Economists were reluctant to give up on their economic models which assume that the economy is only knocked off course by random events. Irrational behavior is not countenanced in their models of rational actors with perfect foresight. They assume all risk can be captured using the mathematics of probability. Unfortunately, that is not what actually happens in real markets. Economists ignored the social scientists and historians who have long understood that the more a quantitative social measure is used as a target the more it will distort and corrupt the social processes it is intended to monitor. That is exactly what happened when granting mortgages to people who could afford them turned into aggressive salesmanship. Jerry Muller, the historian, is quoted as saying '*anything that can be measured and rewarded will be gamed*'[24].

And furthermore, it can become illegal. Weak regulation means that it often goes undetected for too long. According to the UK BBC Panorama TV programme[25] in August 2022, failed investment schemes based in the UK cost over £UK1bn. The TV programme tells the story of three of them Blackmore Bonds—investing in residential property around the UK to build homes and get payback from purchases and rents, Blackmore Global and London Capital & Finance.

The trigger was a change in UK law in 2015. It allowed people to cash in their pensions and invest the money themselves. The politicians argued that it gave people more freedom. Unfortunately, it also made them vulnerable to slick and persuasive marketing by unscrupulous dealers. Blackmore attracted investors by promising particularly good returns of 10% per year. The scheme, known as a mini bond[26], should by law only be sold to experienced and sophisticated investors. Blackmore ignored that law and sold their bonds to inexperienced investors. Many of those investors say that they told Blackmore very clearly that they could not afford to lose their money. They were reassured by the Directors that their money was safe. Blackmore Bond collapsed in 2020. Accounts showed large payments to companies owned by the Directors some overseas in Gibraltar and Isle of Man for fees and expenses. No houses had been built. Over 2,000 investors lost £46m. A web of complex transactions around the scheme made the money difficult to trace. Blackmore's publicity material made the mini-bonds sound like a good investment and in any case the investors were told that their money was protected by an insurance scheme. The insurance scheme, based in Costa Rica, never paid out. Blackmore were accused of using so-called boiler-room techniques[27] for high pressure selling including cold calling. Such techniques for these complex investments are officially banned.

The UK Financial Conduct Agency (FCA) was set up to protect investors in 2013. The FCA were warned several times of boiler room agencies selling toxic and worthless bonds. Direct warnings about the tactics of a company marketing the Blackmore Bond were made to the FCA in 2017 and 2018. In June 2021 questions were raised in the House of Commons regarding the behaviors of the two directors of Blackmore Bonds of whom Peter Grant MP said '*whether the conduct was criminal, civilly unlawful or simply despicable...that to describe it as despicable would be excessively charitable*'[28]. Cross-party MPs, including members of the Treasury Select Committee, have called for an inquiry into the FCA's handling of the Blackmore Bond. The FCA says it shared intelligence about the Blackmore Bond with the City of London Police in 2017 and forced the sales company to shut its website. In 2022, the FCA issued a statement on its website '*We believe this firm may be providing financial services or products in the UK without our authorisatio*n'.

Another earlier case was that of London Capital & Finance (LCF). They issued various bonds to commercial lenders in its own name for up to five years. They offered various rates of interest depending upon the terms of the bond. The scheme collapsed in 2019, leading to 11,600 investors losing £236m. The administrators found that there were a number of highly suspicious transactions with large sums of investors' money ending up in the personal possession of a few people. A large number of borrowers did not appear to have sufficient assets to pay back investors. Investors' money was loaned to a complex web of companies, many of which were controlled by LCF people and a quarter of all the money invested was paid straight to LCF's marketing company[29,30]. The LCF collapse led to an inquiry by former Court of Appeal Judge, Dame Elizabeth Gloster, and some compensation for the victims[31]. The MPs believe the same should happen in the case of Blackmore.

And it had happened before. In 1960 Bernard Madoff founded a Ponzi scheme[32] until his arrest in 2008 when his firm was worth US$64.8bn. In 1999 Jordan Belfort, now known as the 'The Wolf of Wall Street' pleaded guilty to fraud by stock market manipulation through a penny-stock scam. This is where false and misleading statements are used to hype stocks to inflated prices and then sold to make massive profits of many US$ millions. He is reported to have said '*I got greedy.... Greed is not good. Ambition is good, passion is good. Passion prospers*'.[33]

As I write this in late summer 2022 some economists are commenting that the Chinese property boom is a giant Ponzi scheme, bigger than the American banking sector in 2008, which is likely to go bust[34].

The Resentment and the Anger

The picture seems to be all doom and gloom. People have lost trust, if they ever had it, in their political leaders and some businessmen. As John Lanchester points out[35] people are angry about many things—the years of austerity, ever increasing inequality, the seeming immunity of finance companies, the increasing corporate profits, the declining of real pay and, in 2022/23 a cost-of-living crisis that is creating poverty of those in-work. Electorates have turned on politicians in a big way. This sense of a system gone wrong has led to a sense of despair. In recent years political crises across the developed world have led people to vote for easy promises as evidenced by Brexit, Trump, and variously surprising electoral results from Italy, Hungary, Poland, the Czech Republic and elsewhere.

Since the 2008 banking crash we had made very little progress towards reform of the banking system and international finance. It has become clear that bankers were grossly overpaid and that they had incentives for taking risks that paid them huge bonuses when the bets succeeded, but when they went wrong, the losses were paid for by us—taxpayers. New legislation has enforced delays before bonuses can be paid out, and they can be clawed back if things go wrong. But overall remuneration in finance is still extremely high.

Another concern is to separate (casino like) investment banking from retail banking (that serves us all day to day). But there has been no real separation. Instead, we now have a hugely complicated, and technical process of internal ring-fencing inside our banks.

The four biggest UK banks have balance sheets that, combined, are approximately twice as big as the UK economy. The UK financial sector is about 30 times the size of its GDP. History tells us that this kind of undue size and complexity 'incubates' an exploitation of loopholes. It enables very clever, well paid, people to spend all day every day, thinking of ways to get around the rules. Complexity works to their advantage. Banks use borrowed money so that when they make losses, they mainly lose other people's money. Banks should be made safer by making them increase their equity so that they would lose more of their own money before they lost anyone else's. Banks are still too big to fail. Indeed, the problem is worse because the banks that have survived are now even bigger than before. Shadow banking escapes banking rules. These are financial institutions that don't have a formal banking license such as credit card, insurance, and companies such as PayPal.

Ultimately the burdens all seem to fall much more on the poor than on the better-off. Tiny changes in state spending can have direct personal consequences on the poorly paid. Inequality is still growing[36]. But it is not just inequality of income. It is also inequality of wealth, inequality of opportunity, and inequality of access to health care or power. The rise in all of these, leads to a lack of social mobility[36] and a sense in many people of 'us and them'. A sense that there are different rules for insiders with an ever-widening abyss between the people at the top of the system and everyone else.

As I write this, the UK is reeling from the four shocks of the banking collapse of 2008, the short-term consequences of Brexit, the Covid-19 pandemic, and the Russian invasion of Ukraine. It is apparent that even some people in work cannot cope with the rapidly rising cost of living crisis caused by the sudden increase in the international prices of oil and gas. The people of the UK face a winter of hardship. The UK Trades Unions are finding support for strikes to increase wages to counter inflation. The risk is of a wage-inflation spiral. The world of finance seems to be in a bad place in the late summer of 2022.

DEBT

Central to all of these concerns is the role of debt. The 2022 UK cost of living crisis means that many millions of people are facing fuel poverty and debt. Debt is devastating for individuals—and even more so if you have been scammed of your life savings by greedy and unscrupulous individuals' intent on defrauding the systems by exploiting weak regulation. But what of government debt—how do governments around the world legitimately sustain the large annual deficits that add to their massive debt burden?

Let's remind ourselves of the basics. First, we must recall that households as currency users and governments as issuers of fiat currency are very different. To both debt is something that is owed or that one is bound to pay to or perform for another. A debt of £UK 50 is a liability or obligation to pay or render that amount. Some debt is unsafe (bad) because the borrower can't pay it back and could end up bankrupt. On the other hand, safe debt is an amount owed which can be paid back and is good debt if used for a good purpose. Safe debt used for a bad purpose is bad debt. Unfortunately, safe debt can become unsafe through no fault of the borrower or debtor. For example, you might take out a mortgage to buy a home. It is a good debt when you borrow the money, because you are married, you both have jobs with good salaries and every expectation of being able to afford the payments. Then after a few years, the economy goes into recession and one (or both) of you gets laid off. The government response is weak, and the economy remains depressed for years. You drain your savings trying to keep up the payments. But you've got problems if house prices have fallen, and you can't even sell the house to get out of debt. Your last resort is to file for bankruptcy.

Governments with large debt in their fiat currency have never defaulted on repayments. They have safe debt. They are rather like someone with a mortgage who repays the loan plus interest over a number of years and never defaults. Governments with large debts in the currency of another country can have safe debt if their economy is strong. They are rather like someone taking a mortgage with a foreign company in the currency of that foreign country. They can get into unsafe debt if they have insufficient reserves in that foreign currency or the foreign exchange rate worsens to such a degree that they can't pay back what they owe. If government use their safe debt wisely, for example to reduce inequality, then they have good debt. If they use their safe debt unwisely, for example to reward the rich and not the poor, then their debt is bad. In other words, safety is necessary but not sufficient for good.

As Stephanie Kelton[38] points out, there is no shortage of top politicians who haven't understood why government with fiat currency can never run out of money—never go bankrupt. Kelton is a major advocate of MMT (Modern Monetary Theory). She quotes Margaret Thatcher as saying in a 1983 speech[38] '*The state has no source of money, other than the money people earn themselves. If the state wishes to spend more it can only do so by borrowing your savings or taxing, you more*'. She added '*We know that there is no such thing as public money. There is only taxpayer money*'. When Barack Obama was asked in a TV interview, '*At what point do we run out of money*?' He replied[39], '*Well, we are out of money now*'. In effect he said that the USA was broke. Theresa May in the UK said that her government doesn't have a magic money tree[40]. Echoed by US Senator Ted Cruz who said in 2020[41] '*It's not like there's a magic money tree in Washington*'. Even today politicians say we should not leave large debts for our children and grandchildren to pay. Headlines terrify voters when they scream or record indebtedness and looming disaster. More importantly the misunderstandings lead to policies that do not invest sufficient funds to tackle the big problems of climate change and rising inequalities. They lead to policies that are killing to golden goose.

Kelton[42] reports how she asked some US politicians to imagine they were given a magic wand that could eliminate the entire national debt with one flick. They all waved the wand because they wanted to get rid of the debt. Then she asked a seemingly different question—suppose the wand could rid the world of all government securities—would you wave it? They were puzzled but eventually decided they wouldn't wave the wand to do that. She then pointed out that the two questions were simply different versions of the same.

Kelton wants us to stop thinking that selling securities is borrowing and that the securities themselves are debt.

But as I write this in May 2023 there is a crisis in the USA over the so-called debt ceiling. Congress has the ability to set a limit on how much the US government can borrow. Painful negotiation between the political parties requires both the Republicans and the Democrats to change spending policies. If there is no agreement then the risk is that the federal government will default on its loans, investors could lose faith in the US$, with stocks falling and jobs cut. As Stephanie Kelton comments[37] '*to sell the American people on the idea that it would be irresponsible to raise the debt ceiling without negotiating around the budget, Speaker McCarthy invoked every deficit myth in the book. Literally*'. We have already referred to MMT and Stephanie Kelton's myths in earlier chapters and we will look at them again in a moment.

Many economists call for countries to have sustainable levels of government borrowing. It seems to be simple common sense. However, there is no agreed way of defining what this level should be. Being sustainable means being able to meet current and future payment obligations without external financial assistance or default on its loans. Indicators of problems are high ratios of debt to GDP or borrowing to GDP and upward trajectories of interest rates because they affect the cost of borrowing. The International Monetary Fund (IMF) analyses the capacity of a country to finance its policy objectives and debt. It uses a formal framework with a baseline of macroeconomic projections and sensitivity tests. Indicators, such as debt to GDP ratio, foreign currency debt, average years to maturity, etc., are used to assess the vulnerability of the country to a payment crisis[43].

The starting point to getting to grips with national debt is to see that all fiat currency has first to be issued by a government. Then the government takes some of it (say around 90%) back by taxation. That leaves 10% in the private sector.

Typically, the government spends more than it receives in taxes. For example, it wants to provide for the shared public good such as health and welfare, education, transport, and a myriad of other good things. The spending in excess of the income from tax is called deficit spending. To cover that deficit spending it sells an equivalent number of securities on which it pays interest. That is what we call borrowing, and the interest is the cost of borrowing. But it is important to note that the borrowing has been supplied to the private sector by the government's own deficit spending. The government is not selling securities because it needs the money. Its borrowing is just moving around the deficit spending for other reasons as we shall see later. So, in sum, a currency issuer's spending is self-financing except for the interest payments. Indeed, the government could just abandon taxes and borrowing altogether and create the money it wanted to spend. But that would bring at least four sets of problems. First taxes are needed to motivate people to work and produce things in exchange for currency. Second taxes allow the government to alter the distribution of wealth and income. Third taxes enable the government to change behaviors to improve public health (e.g., discourage smoking). Fourth abandoning taxation and government borrowing would cause inflation. This is because boosting spending on the common good would require reducing the spending power of the rest of us. The boost in cash would push up prices.

Critics of MMT say that the cost of borrowing has eventually to be paid for in order to have sustainable levels of debt. At root this means fiscal policies of raising taxes or cutting spending with implications for other political and social goals. Government could also issue money or roll over the cost by issuing or borrowing as the securities come due. But these, say the critics, are inflationary because issuing more money lowers its value against goods and services.

My learning point number 5 is that the deficit of Governments with fiat currency (Government spend > tax) puts money into the private sector.

The deficit creates additional money in the Bank Sector deposit (liability) accounts with the same amount in Reserves (asset) accounts. Reserves are a Banking Sector asset that normally earns no interest. So, when the government offers to sell Treasury Bonds equal to the size of the deficit, the Banks can convert non-income-earning Reserves into income-earning Treasury Bonds. This is why every sale of Treasury Bonds in history has been not merely successful but oversubscribed.

Learning point number 6 follows. It is that the purpose of selling Government Treasury Bonds in fiat currency is not to borrow money from investors rather it is to avoid an overdraft in its account at the Central Bank.

If the Treasury didn't sell its bonds money would still be created because, the Banks's liabilities account would still be Spend minus Tax. The deficit creates money, regardless of whether or not bonds are sold. If no bonds were sold the only flows into and out of the Bank's liabilities account would be Tax minus Spend. With sustained deficits over time, the Treasury's deposit account at the Central Bank would go negative—it would turn into an overdraft account. For individuals, an overdraft attracts a much higher interest rate. If the overdraft gets too large, then bankruptcy looms. But the Treasury owns the Central Bank, so the profits are remitted to the Treasury. In effect the Treasury pays zero interest on its 'debt' to the Central Bank. There are no consequences for the Treasury from having an overdraft at the Central Bank.

This takes us to another *major learning point number 7. Debt need not be a burden to our children and grandchildren.* Again, remember that a government with fiat currency is a currency *issuer* not a currency *user*. The conventional idea that it must balance its budget, like a household or small-sized business is wrong. Today's current expenditure does not have to be offset by current taxes. Future generations will not have to pay for public expenditures from which they didn't benefit. There are four reasons. First, unlike a household, government spending and income are not independent of one another. Investments in productive income-generating activities are multiplied by factors, called multipliers, that, when changed, cause changes in other related economic variables. The multiplier causes gains in total output to be greater than the change in spending that caused it. These can boost growth, employment, and tax revenue. Borrowing for green investments, for example, have a number of positive economic multipliers and may well pay for themselves. Second, what is important is the affordability of servicing debt. The ability to pay off interest charges. Public finances are sustainable indefinitely when debt servicing costs are equal to or below the growth rate of the economy. If interest rates are lower than the increase in national income, the relative debt burden will shrink over time on its own. Third, there are government assets which will be left to future generations. These include functioning health and social services (NHS in UK) and healthy ecosystems. Fourth, government borrowing doesn't have to be funded by taxation. It can be financed as government borrowing. This is how the UK dealt with the Covid-19 crisis, helped by the central Bank of England to control the level of interest that the government has to pay on its debt.

This leads us to *learning point number 8 that interest on Treasury Bonds can be paid by the Treasury borrowing from its own Central Bank.*

The Treasury adds to a loan on which it pays zero interest. The government is in debt to itself, and the interest it pays on the bonds. The interest payments are another way in which the government creates fiat money in the private sector by adding to the equity of the Banks.

But, having said all of that, there are constraints. As we have said inflation is perhaps the most important one. Monetary policy cannot deal with constraints in real resources. Borrowing can be a burden for future generations if it results in an inefficient allocation of those resources. A good example is government borrowing to finance tax cuts as proposed by UK Prime Minister Liz Truss in her Autumn minibudget 2022. This is safe debt used for a bad purpose and resulted in her record-breaking quick downfall. What is more, tax cuts for the rich increase inequality. Fiscal responsibility to borrow and tax strategically is a must.

BOOKKEEPING

One perhaps unlikely key to a way of understanding these points was invented long ago by the Italian traders of the 13th and 14th century. It is double entry bookkeeping. This kind of bookkeeping is required to create balance sheets which are important not just as a snapshot at a point in time of a company's accounts but because they are essential to understanding the flow of money between all financial entities from individuals to governments.

If you want to understand this in more detail, please refer to the Appendix—The Flow of Money. There we look at a stock-flow model and the use of Godley tables by economist Steve Keen. Every financial institution whether individuals, companies or governments has assets and liabilities. You, like many reading this book, probably do not write balance sheets for your personal finances—very few people do. But it can be instructive to do so. For example, your assets are any items of economic value that belong to you. These may include your house (in total if you own it or that part of your mortgage you have paid), car, jewelry, your bank and building society savings and chequing (or current) accounts if they are positive (including your salary or wage income if paid into your bank), life insurance policies, annuities, investments such as shares and bonds, antiques, art collections, electronic equipment and any other items that have cash or market value. Your liabilities will be financial. These include everything that you owe such as house mortgage amounts yet to be paid, other loans e.g., for a car or to purchase equity or any other product such as a TV or computer, credit card debt, monthly bills on rent, utilities such as water, electricity and gas, insurance payments, hospital bills, etc.

Government balance sheets of course are vastly more detailed and complicated than personal ones—but the principles are the same. Data may be grouped under various headers such as by sector (public or private, financial, or non-financial), asset (dwellings, other buildings and structures, machinery, equipment and weapons, intellectual property, cultivated biological resources, minerals, artistic originals, money, deposit, debt securities, loans, equity, insurance pensions, derivatives, stock options, and accounts payable). Liabilities include special drawing rights, currency and deposits, debt securities, loans, equity, insurance schemes, derivatives, and stock options. For the UK in December 2021, the assets were estimated as £UK 46.9 trillion and the liabilities were UK£36.2 trillion. All of this information is published online but a simple version is Figure 4.1. The difference between total asset and total liabilities is called net worth. The net worth of the UK in 2021 was therefore UK£ (46.89–36.24) trillion which rounds to £UK10.66 trillion[44].

Modern Monetary Theory (MMT)

I have mentioned Modern Monetary Theory (MMT) many times. In recent years, especially after the publication of Stephanie Kelton's book '*The Deficit Myth*'[45], there has been much discussion about this new theory built on some old ideas. It is a controversial and hot topic amongst economists. The theory is another example something simple and complex at the same time. The controversy amongst economists' centres around the implications of three statements, (a) that currency issuing countries are financially unconstrained, (b) taxes are not needed to finance government spending, and (c) taxes reduce the supply of money by taking money out of the economy after the government has spent it.

The statements on their own seem relatively simple but the debate is about their possible complex consequences. Some economists interpret MMT as advocating monetization—the turning of all things into revenue generating activities, services, or assets. This is where it all gets more complex. For example, as we have seen, a central bank monetizes its government debt by converting securities into credit or cash. In other words, they say MMT advocates that there are no limits to credit. At times of national emergency such as war or pandemics the government spends first to deal with the situation and sorts the consequences out later. These MMT economists say that governments can do this as a matter of course.

All economists agree that there must be constraints. If a government spends vast amounts of money or cuts taxes dramatically, regardless of supply, demand, and net exports, then we can get massive inflation. A country with few natural resources, little industry and human capital cannot just create money to generate prosperity. It may have the domestic power to do so internally but it soon gets into trouble when trading internationally. A country can't make other countries accept their currency for goods and services, machinery, and talent if it has little intrinsic economic value. Countries such as Argentina, Brazil, Weimar Germany, Zimbabwe and most recently Sri Lanka (some of which we will come to later) have tried this and created hyper-inflation.

We have highlighted earlier that currency issuing countries can create money. There is a magic money tree, but it is a constrained one. The way it works is that governments spend money into the economy and take back taxes. When tax does not cover the spending then the government borrows money out of the economy by issuing bonds. Central banks buy the bonds directly from the government (debt monetization) via the primary market as mentioned earlier in this chapter. Then other banks buy the bonds from the primary dealers in a secondary market.

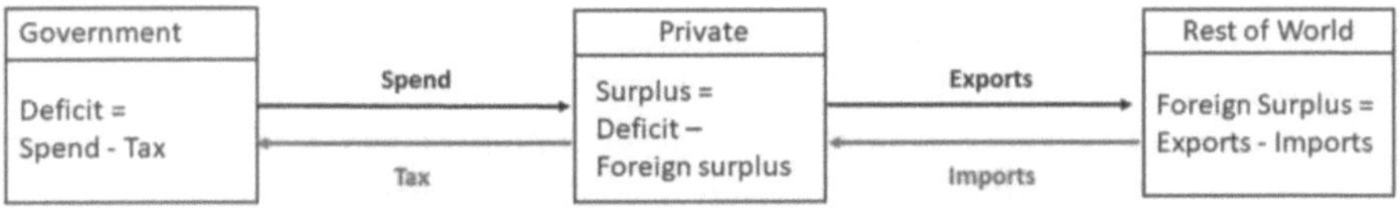

Figure 6.1 An overview of flow of money between the three main sectors of an economy[46]

The key to understanding MMT is to know that every country has three key accounts which together are an overview financial record for that country's currency (Figure 6.1). The first is called the fiscal account. It is between the government (so-called public-sector) and the rest of us (the private sector). MMT economists say that when a government has a fiscal deficit (as often is the case) it can be financed by the independent central bank that it owns. In other words, the government always should spend more than it receives (possibly around 90% as taxes) because that puts money into the private sector. Some non MMT economists claim that in the USA during the Presidency of Bill Clinton the government was in surplus and that was the crowning achievement of modern Democratic governance—but the real story is more complex according to Steiner[47]. The second account is the private sector account itself (the central box in Figure 6.1). This records the balance between households and businesses. To be sustainable that balance must be positive since the private sector must always have more income than it spends if it is to stay solvent over the long term. It can be balanced by a positive third account—the current account. This is between the country and the rest of the world. If imports are less than exports the country is in deficit with the rest of the world and the rest of the world is in surplus with the country. Clearly the reverse applies if imports exceed exports. The three balances account for the country's net financial assets over a given time period and 'as a matter of accounting must always sum to zero'. Stephanie Kelton writes[48] that she learnt this from British economist Wynne Godley. Godley said to her '*Everything must come from somewhere and then go somewhere*'. For every payment that flows out of an account a payment of equal size must go into another account. This simple fact happens every day. For example, as the government makes a contract with a construction company to design and build a bridge then currency will flow from the government sector into the private sector. The companies and the people working on creating the bridge will pay taxes back to the government. By taking taxes the government is removing money out of the economy having first put some money in by paying the bridge builders. As the government spends money into the economy and taxes around 90% of it back then that money circulates around the country's private sector as people pay for goods and services from houses to haircuts.

But what about imported goods? If imports are greater than exports (a trade deficit on the current account) then the government's fiscal deficit is exactly balanced by surpluses in the private sector account and the current account.

For example, if a country has a fiscal balance of –5% of GDP (the negative indicates a deficit) and the country's current account balance is +1% GDP (the rest of the world has a trading surplus with the country and the country has a trading deficit with the rest of the world) then the private-sector balance must be +4% (positive surplus) because (–5% + 4% + 1% = 0).

This point is crucially important to the operation of MMT. This is because if the country creates and spends money entirely domestically then all is well. But, if it requires, as it does, the currency to be accepted as internationally stable then it must have an appropriate level for its foreign trading balance. The level of that balance depends crucially on the foreign exchange rate—a rate that a country can influence but not totally control. If the economy of a country is in trouble such that any money it creates cannot be backed up by real goods and services, then that money is effectively funding spending which is beyond that country's means. The fiat is destroyed and meaningless. As the foreign traders realise this then the money will rapidly become worthless, and the country's foreign exchange rate will drop. Hyperinflation will follow—though there may be a lag time before this bites.

The factors that influence foreign exchange rates are many. They include general economic health (often measured as a growth in GDP), levels of inflation, interest rates, and public debt, political stability, the current account trade balance, and the confidence level of speculators on the foreign exchange market. Inflation occurs when domestic demand is ahead of supply, and this undermines the foreign exchange rate. Inflation can be imported if imports are more expensive than outputs. This is happening in the UK as I write this in the summer of 2022 due to a hike in energy prices caused by the Russian invasion of Ukraine. If the imported inflation gets embedded in a nation's economy as it seems to have done in 2023, then the consequences depend crucially on the underlying strength of that economy. If the health of an economy is weak then government funding of imported inflation will result in lower foreign exchange rates and a collapse of the kind, we have seen in Sri Lanka (see later). If the underlying economy is strong enough, then governments should be able to borrow to fund the trade imbalance with only a limited impact on the foreign exchange rates. The inflation must then be tackled by increasing interest rates although the government buying of bonds by the central bank will tend to counter this.

Godley Tables

Wynne Godley (1926–2010) was a British economist who saw the importance of integrating the modeling of all the stocks and flows in an economy. Steve Keen (1953–) has developed a macroeconomic simulation model based on Godley tables. He calls them Godley tables in tribute to Wynne Godley's insight. He calls his developing software *Minsky*[49] in tribute to the economist Herman Minsky (1919–1996). The tables are a version of a balance sheet for each financial entity in the model. They include assets, liabilities, and net worth or equity. Each column of a Godley table contains the accounting record of a financial entity recorded as an asset or liability or the net worth. It uses double entry bookkeeping so every asset in one table must be a liability in another table. Each row in a Godley table has two entries recording a specific type of transaction at a macroeconomic level. Examples are the payment of wages, interest, or the creation of new debt. The entries in a row must sum to zero. Godley tables are, effectively, connected balance sheets that change in time. A stock-flow model like Keen's *Minsky* software can model them as they exchange data that can be simulated through time in chosen time steps. The idea is remarkably simple and yet powerful. For more detail see the Appendix: The Flow of Money. Hyman Minsky was an American economist whose ideas were largely ignored until the subprime banking collapse of 2008 which the magazine The New Yorker called a '*Minsky Moment*'. Minsky argued that private accumulation of debt was a big contributary factor to a financial crisis. He described the three types of borrowers as hedge, speculative, and Ponzi. A hedge borrower makes debt repayments from invested current cash flows. A speculative borrower services the debt interest but regularly re-borrows the amount of the loan. A Ponzi borrower believes he can finance his loan through the appreciation of the value of the asset.

The Weimar Republic, Argentina, and the Sri Lankan Crisis

What is the history and consequences of financial mismanagement? We'll examine briefly just three. The commonest outcome is hyperinflation with countries needing international bailouts to recover.

One of the most quoted examples, by opponents of MMT, is hyperinflation in the Weimar Republic. In the early post WWI years, the German Reich (1918–1933) created (in other words printed) currency to pay its debts. By 1923 it could no longer afford the reparation payments required by the Treaty of Versailles and the government defaulted. French and Belgian troops occupied the Ruhr where most industry was located and took control of manufacturing and mining companies. The workers went on strike. They were paid by the money created by the state. Hyperinflation followed because Germany had few goods to trade, and the printed currency was worthless outside Germany. In 1924 American banks lent money to German banks with German assets as collateral to help it pay reparations. The German railways, the National Bank and many industries were therefore mortgaged as securities for a stable currency and the loans.

Argentina became independent from Spain in 1816. Since then, it has defaulted on its debt nine times[50]. Inflation reached an astounding maximum of 5,000% in 1976. Argentina has richly fertile pampas land. Economic growth in the early 20th century was good so that by 1913 Argentina was world's 10th wealthiest state per capita. But it had political problems. In the 1930s a military junta took power with almost inevitable political instability. The government wanted Argentina to become industrially self-sufficient. Investment was diverted away from agriculture. Later in 1946 the dictator Juan Peron was popular amongst the Argentine working class by promising land, higher wages, and social security. He gradually increased state control with higher government spending. Populist protectionist policies and inefficient production created chronic inflation. After the fall of Peron in 1955 discontent with economic policies led to a wage-price spiral in the 1970s. Peron returned to power in 1973 and expanded the supply of money. Uncontrolled inflation ensued. Foreign debt rose to three quarters of GDP by the late 1980s. The government pegged the peso to the US$ but a recession caused a default at the turn of the 21st century and again in 2014 and 2020.

Before Covid-19 struck Sri Lanka was paying down its debt with revenue from tourism and new loans. The pandemic severely affected tourism and its trade balance became such that the country could not pay for essential imports or its debt repayments. The government banned herbicides and fertilizers to save US$ 400 m per annum on imports with a move to organic agriculture over 10 years. The farmers' protested and were soon joined by other workers. The trade balance got so bad that the government couldn't pay for imports with consequent shortages in essentials such as medicines. The government created new money to pay for goods and inflation grew to 50%. When people could no longer pay for cooking gas and fuel the government said it would provide fuel for essential services only. Schools were shut and workers ordered to stay at home. On paper the country seemed to be developing well with a rising GDP. But the growth came from external money rather than Sri Lankan goods and services. Short term loans from China with high interest resulted in a debt of between $5–$10 bn with an overall debt of $51 bn. Sri Lanka needs large-scale, long-term economic restructuring but to get international help the country will need to be politically stable. It must negotiate new terms with the IMF with perhaps new lower interest rates and longer-term loans. The IMF will not lend Sri Lanka money simply to pay off its debt to China or anyone else. The IMF is also likely to expect that the Sri Lankan rupee LKR floats free of its peg to the US$ together with other austerities.

What do these examples teach us? There are at least five lessons and probably many more. First political stability is vital. It is the basis of the fiat. Investors, understandably, do not like uncertainty. Second the underlying health of an economy must be sound. Labour problems, lack of government investment and weak economic policies all contribute. Third, central banks must be aware of the value of the money they create. If money has no intrinsic economic value—that means it cannot be used to obtain goods and services—then it rapidly becomes worthless. This could happen if the money created is more than that required in an economy even if it was created for some laudable reason such as encouraging banks to lend or paying increased wages. Fourth, there should be economic growth—if there isn't then business raise prices to boost profits and stay in business. Finally, and perhaps most importantly, people in both public and private sectors need to have confidence in the economic system. They will gain that by examining various indicators such as the levels of economic activity and the behaviors of market traders—especially on the foreign exchanges. A weakness in one currency against another (e.g., the Sri Lankan rupee LKR against the US$) has several implications. The first is that it ought to make Sri Lankan exports to the USA cheaper. The second is that it risks further pushing up costs for Sri Lankan companies in buying dollar-denominated commodities—the most important of which, of course, is oil. The third is that Sri Lanka will become attractive to tourists.

Money is particularly important when you don't have any. Banks advise us to 'make our money work for us' by using our money to make more money. To take control of our finances to improve the stability and security of our income and expenditure and be financially independent to build wealth by investing. There are different forms of money. Money and wealth are different, and both are also very simple and complex. At its simplest money is a medium of exchange and wealth is net worth. Both become complex when we try to manage them through fiscal and monetary policies. These policies are very blunt instruments, and we need better automatic stabilizers. The policies require a very great deal of careful interpretation and the building of confidence in all stakeholders including the markets. Governments set fiscal policy with only three main instruments, tax, borrowing and spending. But what the politicians say is as, if not more, important in building confidence. Central banks set monetary policies with only two instruments interest rates and the setting of minimum bank reserves. Likewise, what the Governor or President of the central bank says is also important in how investors react. The reason why economic forecasting is so difficult and uncertain is because the economy depends on the collective opinions and intentions of millions of people. The amount of money in circulation is called the money supply. It is not widely appreciated that when banks make loans, they increase the money supply. Central banks also use what economists call open market operations to control the supply of money. These are the buying and selling of government securities (bonds or IOUs). They buy using newly created money that is deposited in a bank's account with a central bank. The money is created out of 'thin air'. But not without constraints—chiefly the risk of inflation. Government bonds are the most secure forms of borrowing for the private investor because governments with fiat currency will never default.

Economic cycles that affect the value of money are inevitable—expansions followed by contractions or recessions. History tells us they are recurrent but not regularly periodic and therefore hard to predict. They arise from changes in production and trade. Hedging is a way that investors try to protect their assets. They 'hedge' one investment by making another—it is a complex and risky activity. Arbitrage is about finding ways of removing risk so that whether you bet that the market will go up or go down you will still make a profit. It is the simultaneous purchase and sale of an asset in different markets to exploit tiny differences in their prices. Through the kinds of manipulative trading of financial assets rather than goods and services, many people have lost trust and are angry. They have endured years of austerity, ever increasing inequality, the seeming immunity of finance companies, the increasing corporate profits, the declining of real pay and, in 2022 a cost-of-living crisis that is creating poverty of those in-work. Partly as a result electorates have turned on politicians in a big way and are listening to the simplistic policies of populists and extremists. Populism thrives when leaders lose sight of the concerns of their own citizens.

Debt is devastating for individuals. How do governments around the world legitimately sustain the large annual deficits that add to their massive debt burden? This is part of a major misunderstanding. Government debt need not be a burden to our children and grandchildren. First, unlike a household, government spending and income are not independent of one another. Second, what is important is the affordability of servicing a level of debt. Third, government spending leads to assets which will be left to future generations. Fourth, government borrowing doesn't have to be funded by taxation. Modern Monetary Theory (MMT) has been, and still is, controversial. The key to understanding MMT is to know that every country has three key accounts which together are an overview financial record for that country's currency. The theory is another example of something that is simple and complex at the same time. The three balances account for the country's net financial assets over a given time period and as a matter of accounting must always sum to zero. By using double entry bookkeeping to create balance sheets which are essential to understanding the flow of money between all financial entities from individuals to governments. As Wynne Godley said, '*Everything must come from somewhere and then go somewhere*'. The controversy amongst economists, centres around the implications of three statements, (a) that currency issuing countries are financially unconstrained, (b) taxes are not needed to finance government spending, and (c) taxes reduce the supply of money by taking money out of the economy after the government has spent it – that is to manage demand and match it with supply.

The examples of the Weimar Republic, Argentina and more recently, Sri Lanka show what can go wrong is real constraints aren't considered properly. But most important of all is that misunderstandings about debt in fiat currency government policies lead to severe under investment in tackling the big problems of climate change and rising inequalities.

There are at least five lessons that we can learn, and possibly many more. First political stability is vital. Investors, understandably, do not like political uncertainty. Second the underlying health of an economy must be sound. Labour problems, lack of government investment and weak economic policies all contribute. Third, central banks must be aware of the value of the money they create. If money has no intrinsic economic value—that means it cannot be used to obtain goods and services—then it rapidly becomes worthless. This could happen if the money created is more than that required in an economy—even if it was created for some laudable reason such as encouraging banks to lend or paying increased wages. Fourth, there should be economic growth—if there isn't then business raise prices to boost profits and stay in business. Finally, and perhaps most importantly, people in both public and private sectors need to have confidence in the economic system. It is misunderstandings such as these which are killing the golden goose.

End Notes

1. See Aesop's Fables at https://read.gov/aesop/001.html
2. Chancellor, E. 2022. The Price of Time. Penguin Books, UK loc 1127]
3. *Ibid.*
4. *Ibid.*
5. *Ibid.*
6. Kelton, S. 2022. Policymaking in a Pan(demic). The Lens. See https://stephaniekelton.substack.com/p/policymaking-in-a-pandemic?utm_source=substack&utm_medium=email
7. See Chapter 2 and End Note 13. L. Randall Wray. 1997. Government as Employer of Last Resort: Full Employment Without Inflation. Levy Institute Working Paper No 213 at https://www.levyinstitute.org/pubs/wp213.pdf. Also see overview at https://activistmmt.org/job-guarantee/. A jobs guarantee wage is paid only to people working in a jobs guarantee scheme. It is an example of an automatic stabilizer. The more people on the scheme the more the government spends on wages. When those people move to private sector jobs then government spending goes down. In other words, government spending goes up when the economy is down, and spending goes down when the economy is up. Since it is carefully targeted and automatically self-adjusts based upon need, there is no requirement to correct any overspending via taxation. Tax rates are unaffected.
8. McLeay, M., A. Radia and R. Thomas. 2014. Money creation in the modern economy. Bank of England Quarterly Bulletin Q1. https://www.bankofengland.co.uk/-/media/boe/files/quarterly-bulletin/2014/money-creation-in-the-modern-economy.pdf
9. Lanchester, J. 2018. After the Fall. London Review of Books. See https://www.lrb.co.uk/the-paper/v40/n13/john-lanchester/after-the-fall
10. Fischer, B. and S. Myron. 1973. The Pricing of Options and Corporate Liabilities. The Journal of Political Economy. (81)3: 637–654. (May–Jun. 1973). The University of Chicago Press. See http://www.jstor.org/stable/1831029.
11. Lanchester, J. 2010. Whoops. Penguin Books, UK.
12. Investopedia. 2020. Flash Crash. See https://www.investopedia.com/terms/f/flash-crash.asp
13. *Ibid* [Lanchester].
14. Conforming loans are standard mortgages.
15. Li, D.X. 2000. On default correlation: a copula function approach. The Risk Metrics Group Working Paper No. 99–07. http://www.cyrusfarivar.com/docs/li.defaultcorrelation.pdf#:~:text=David%20X.%20Li%20April%202000%20Abstract%20This%20paper,as%20the%20correlation%20coef%EF%AC%81cient%20between%20their%20survival%20times.
16. Correlation is a mathematical way of expressing a *linear* interdependence between two or more random variables. Note the word linear. Two variables describing a circle would have zero correlation. Note also one common error is that of ascribing causation to correlation.
17. A random variable is a mathematical variable that depends on random events. A mathematical variable is a quantity that can change. It is usually represented by letters such as GDP for gross Domestic Product. Randomness is an apparent lack of a pattern in a set of variables or events. Clearly that apparent lack could be because one has not

tested for a sufficient number of possible patterns. As this is infinite then ascribing randomness to a variable is always tentative.

18. A copula function links a single random variable to a multivariable joint probability distribution. They are used in probability theory to model the dependence between random variables.
19. *Ibid* [Lanchester].
20. *Ibid* [Chancellor].
21. King, M. and J. Kay. 2020. The End of Alchemy. Abacus/The Bridge Street Press, UK.
22. *Ibid* [Chancellor].
23. Bunn, P., P. Mizen and P. Smietanka. 2018. Growing Pension deficits and the Expenditure Decisions of UK Companies. BoE Staff Working Paper, London. See https://www.bankofengland.co.uk/-/media/boe/files/working-paper/2018/growing-pension-deficits-and-the-expenditure-decisions-of-uk-companies.pdf
24. Muller, J. 2017. The Tyranny of Metrics. Princeton University Press, USA. See https://www.goodreads.com/work/quotes/58414707-the-tyranny-of-metrics#:~:text=%E2%80%9CThere%20are%20things%20that%20can,be%20greater%20than%20the%20benefits
25. BBC TV (August 2022). *Panorama*. London. See www.bbc.co.uk/iplayer/episode/m001b7jh
26. Mini-bonds are a direct investment in a company for a fixed return over a set period—they cannot be traded like bonds.
27. Boiler room techniques are used by call centre to sell dubious investments by telephone with dishonest sales tactics.
28. See Hansard at https://hansard.parliament.uk/commons/2021-06-30/debates/3BB3C55C-491F-4CBB-953C-AA80C097E1C9/FinancialConductAuthorityAndBlackmoreBondPlc.
29. See https://www.bbc.co.uk/news/business-62504445
30. See BBC News 27 March 2019. https://www.bbc.co.uk/news/uk-england-47713230
31. Gloster, E. 2020. Report of the Independent Investigation into the Financial Conduct Authority's Regulation of London Capital & Finance plc. See https://assets.publishing.service.gov.uk/government/uploads/system/uploads/attachment_data/file/945247/Gloster_Report_FINAL.pdf
32. A Ponzi scheme is a fraud that attracts investors and uses their money to pay profits to earlier investors.
33. See https://en.wikipedia.org/wiki/Jordan_Belfort#Restitution
34. See ref https://youtu.be/JJeWBxhXWGM
35. *Ibid* [Lanchester, Reference 6].
36. The current overall level of inequality in the USA is approaching the extreme level prior to the Great Depression. The lower-tail inequality rose sharply in the 1980s and contracted somewhat thereafter, while upper-tail inequality has increased steadily since 1980. See *20 Facts About U.S. Inequality that Everyone Should Know*, Stanford Center on Poverty & Inequality (2011) at https://inequality.stanford.edu/publications/20-facts-about-us-inequality-everyone-should-know. The so-called Great Gatsby Curve is a graph, first developed by Alan Krueger in 2012, that shows the relationship between income inequality in a given country and the potential for its citizens to achieve upward mobility. It seems to demonstrate a strong positive correlation between inequality and a lack of social mobility. See Kurt, D. 2022. The Great Gadsby Curve. Investopedia.

at https://www.investopedia.com/the-great-gatsby-curve-5205821. Steven Pearlstein points out that most of us would accept that freedom to exchange goods, services and money is fair and just. Likewise, people are entitled to own and keep what they produce. But is it right that rock stars earn more in a month than school music teachers earn in a lifetime? Such skewed distributions seem at odds with our moral intuitions. The super rich are pulling away from everyone else and thought of by many as not just economically necessary bit morally acceptable. Pearlstein, S. 2018. Moral Capitalism; Why Fairness Won't Make Us Poor. St Martin's Press, New York.

37. See also Chapter 1. Kelton, S. 2021. The Deficit Myth. John Murray, USA. See also Kelton's blog 'The Lens' called 'The Deficit Myth is Hanging On for Dear Life' May 30th 2023 at https://stephaniekelton.substack.com/p/the-deficit-myth-is-hanging-on-for?utm_source=post-email-title&publication_id=423556&post_id=124412934&isFreemail=true&utm_medium=email
38. Margaret, T. 1983. Speech to Conservative Party Conference. Winter Gardens, Blackpool, UK. https://www.margaretthatcher.org/document/105454
39. Weisenthal, J. 2009. Obama: The US Government is Broke! Business Insider, May 24. See https://www.businessinsider.com/obama-the-us-government-is-broke-2009-5?r=US&IR=T
40. Dearden, L. 2017. Theresa May prompts anger after telling nurse who hasn't had pay rise for eight years: '*There's no magic money tree*', Independent (London) June 3. See https://www.independent.co.uk/news/uk/politics/theresa-may-nurse-magic-money-tree-bbcqt-question-time-pay-rise-eight-years-election-latest-a7770576.html
41. Twitter, 2020, May 6. See https://twitter.com/SquawkCNBC status/1258024345395441664?s=20
42. *Ibid* [Kelton].
43. IMF International Monetary Fund (IMF). 2011. Modernizing the Framework for Fiscal Policy and Public Debt Sustainability Analysis. See https://www.imf.org/external/np/pp/eng/2011/080511.pdf
44. See ref UK Office for National Statistics National Balance Sheet estimates for the UK: 2021. December 2021. https://www.ons.gov.uk/economy/nationalaccounts/uksectoraccounts/bulletins/nationalbalancesheet/2021#main-points
45. *Ibid* [Kelton] Figure 6.1 is adapted from Exhibit 9 of The Deficit Myth.
46. *Ibid* [Kelton].
47. Steiner, C. 2007. The Myth of the Clinton Surplus. See http://www.craigsteiner.us/articles/16. Steiner says Washington doublespeak and political smoke and mirrors concealed what happened. The US national debt is made up of public debt and intragovernmental holdings. The public debt is debt held by the public, normally including things such as treasury bills, savings bonds, and other instruments the public can purchase from the government. Intragovernmental holdings, on the other hand, is when the government borrows money from itself--mostly borrowing money from social security. Whilst public debt went down from 1998–2000, the intragovernmental holdings went up by a far greater amount. That meant that the total national debt (public debt + intragovernmental holdings) went up.
48. *Ibid* [Kelton].
49. See also Chapter 1. Keen. 2021. The New Economics. Polity Press, London. For Minsky Manual. See https://www.patreon.com/posts/shorter-minsky-71143934?utm_medium=clipboard_copy&utm_source=copyLink&utm_campaign=postshare_creator
50. See https://en.wikipedia.org/wiki/Argentina

Chapter 7

Is the Genie Already Out of the Bottle

Emerging as an Increasing Number of Extreme Weather Events

CRISES ARE OPPORTUNITIES

John F Kennedy[1]: *When written in Chinese, the word 'crisis' is composed of two characters. One represents danger and the other represents opportunity.*

Joseph Stiglitz[2]: *"The world is facing three existential crises: a climate crisis, an inequality crisis and a crisis in democracy".*

The Polar Bears

Polar bears overran the Siberian town Belushya Guba in 2019[3]*. The population of the town is around 2500 people. It lies on the Siberian Arctic Island of Novaya Zemlaya. 52 polar bears invaded the town in search of food with much disruption and concern. Polar bears are not cuddly—they are predators and not afraid of humans. Attacks are rare but if they do attack then they can be fatal for both. The increasing loss of sea ice and hunting grounds means they must move to where there can find food. Waste found in human settlements are attractive because scavenging is less energy-intensive than hunting for seals and offers the bears a more secure and abundant food source. The human response was to send in experts to drive the bears away and even cull them. Despite warnings from ecologists that polar bears will be pushed to extinction by climate change, all we seem to be able to do is to watch this slow-motion extinction unfold.*

Common Sense is not Common

Enjoy the ride. Life is short—learn from the past, enjoy the present, anticipate the future.

Everything exists in relationship to everything else—nothing exists in isolation.

IS THE CLIMATE CHANGE GENIE ALREADY OUT OF THE WORLD'S BOTTLE?

A genie is a demon. The story goes that a poor fisherman goes fishing but only catches an old bottle. When he opens the bottle, he releases a big, bad genie.

We humans are the poor fishermen who are opening the bottle that will release the climate change genie. The cap is not fully off yet, but the genie is slowly emerging—he is not totally out of the bottle, but he is on the way unless we push him back firmly. We will only know for certain where he is in 2023 with historical hindsight—but then it may be too late.

In the meantime, we still have hope that we can push him back into the bottle and seal the cap. But we can only do that by everyone working together. Currently that is not happening.

How do we convince each other, the world over, that the genie is our biggest common enemy ever?

One thing we know from history is that people come together in times of war. Defeating a common enemy unites us.

At the start of World War 2 things looked bleak, but by working together, and with much loss of life and resources, and much suffering, the common good prevailed.

Unfortunately, the climate genie is even more deadly than Hitler. What is more he is more hidden and is creeping up on us so slowly that there are people who still deny he exists.

The climate genie has no respect for race, creed, wealth or any other of the many ways we find to differentiate ourselves from each other. He is a common enemy for all living things on planet Earth. But he is incipient, insidious, hidden, invisible and incredibly dangerous and what is more he is emerging at an ever-increasing pace. He is already having disastrous effects on the natural world. We are seeing massive loss of life and resources through floods, fires, famine, melting ice and other major changes to the environment. Let's be clear the immediate threat is to life on earth—the planet will survive, and Earth will continue to spin lifelessly through space around the sun.

The question for this final chapter could not be more important.

'Are we up to challenging the climate genie?'

WHAT NEEDS TO BE DONE TO PREPARE US FOR A CHANGING CLIMATE

The list is long. Here are twelve important suggestions based on the five categories used by the IPCC.

Energy Supply

First and foremost, governments must require that all fossil fuels are kept in the ground. All countries should move away from using fossil fuels as soon as possible.

Second, governments, investors and companies must invest more in renewable energy including solar, wind, wave, tidal and geothermal power. They should require that the access and stability of energy supplies is resilient. They should explore the capturing and storage of carbon (CCS). We should all improve the efficiency of our use of water.

Land, Water, Food

Third, we should all reduce meat and dairy consumption. Farmers should provide more plant-based products, improve livestock systems and the management of croplands.

Fourth, governments and others should improve biodiversity, for example by reforestation by planting trees in the right places and giving land back to nature through 'rewilding'. Protect forests like the Amazon by stopping the cutting down of trees for farming. Governments should stop this exploitation through stronger laws.

Fifth, international agreements should protect the oceans by preventing overfishing and stop deep sea mining and drilling for oil and gas drilling. Implement integrated policies for coastal management as sea levels rise. Expand marine protected areas (MPAs) as geographical zones for safeguarding biodiversity and maintaining marine ecosystem health and the supply of ecosystem services.

Infrastructure

Sixth, reduce the use of plastics. Governments should legislate, companies should reduce its use in packaging and individuals should not buy plastic products whenever possible.

Seventh, we should all make transport sustainable by reducing our use of cars use, switching to electric vehicles and minimising plane travel. Governments should improve public transport and access to cycling lanes.

Eighth, we should all insulate our homes and move away from oil or gas boilers to green hydrogen, heat pumps or other renewable energy sources. Governments should invest and actively attract private investment to make these available and affordable.

Health

Ninth, enhance our health and social care serves worldwide. Improve nutrition and diets and poverty levels. Share best practice across the entire world.

Society, Economy, and Lifestyle

Tenth, reduce consumption by changing attitudes towards transport, fashion, food, and other lifestyle choices that cause over and unnecessary production.

Eleventh, move from the consumer story to the citizen story. Evolve from our current form of capitalism based on fear and greed to one of principled capitalism based on a renewed set of moral virtues that become social norms.

Twelfth, improve our understanding of risk and resilience, improve our disaster management systems as the number of extreme weather events increase. Learn how to learn together to build trust. Learn to accept ways in which other people live and govern with which we might disagree. Learn to consider that those differences are not worth killing each other for.

For even more detailed suggestions go to Paul Hawken's 2021 book 'Regeneration', and associated website[4]. He wants to weave '*justice, climate, biodiversity, and human dignity into a seamless tapestry of action, policy, and transformation that can end the climate crisis in one generation*'. He suggests that whether we are individual, group, or institutional, we should make a punch list of actions that we can, want to and will do. For example, (a) stop using single-use plastics, (b) purchase renewable energy credits, (c) set a budget for clothes, (d) engage with a citizen's climate action group, my city council to advocate urban green space, (e) start or join a monthly climate and environmental justice reading group. On his website Hawken points out that following the Paris Agreement in 2015, the world's sixty biggest banks have invested $3.8 trillion in fossil fuels and G20 countries have provided more than $3.3 trillion in subsidies for fossil fuels. He is correct in saying that we are entrenched in a banking system that is actively aiding in the destruction of our shared home. He makes several suggestions for change such as asking any companies you are affiliated with to divest from fossil fuel. If necessary, switch your bank accounts, investments, and pension funds. If you are a shareholder, put pressure on the board to have strong climate priorities. Companies should finance regenerative projects using green bonds. Countries should set up green banks to mobilize large amounts of capital for a climate transition.

WHAT CAN I DO?

It is natural for any one individual to feel helpless in the face of such a challenge. Hawken's punch lists are a good place to start. But to integrate them as never before we need some kind of 'cunning plan' as beloved by Baldrick in BBC TV series Blackadder. Movements like Hawken's plan for regeneration and Alexander's plan for the citizens story are excellent initiatives. Deep in our plan should lie the word ***ingenuity.*** We need to encourage and nurture ingenuity in everyone on the planet—whatever the level they operate from labourer to CEO. One technical example is a company to produce 'green hydrogen' from solar energy. It is being developed through an ingenious combination of a photovoltaic device with an electrolyser device[5]. The photovoltaic device absorbs sunlight to generate the electricity that splits water to create hydrogen. Green hydrogen is at the centre of Australia's clean-economy growth plan passed in its Climate Change Act 2022[6].

Ingenuity is not just restricted to scientist and engineers—we all have a capacity for it. For example, individual ingenuity is to act in a way that is new to you. You may never have lobbied your local politicians. You can change that by lobbying them about the need for investment in tacking the impacts of climate change—make your voice heard and use your vote as part of your punch list. Other are to change your lifestyle by eating less meat and dairy products, cut back on flying, use your car less, reduce your use of energy, respect and protect green space, invest your money wisely, and minimise waste.

We need individual and collective ingenuity to begin to address the complexities of extreme poverty, rising inequality, wars over territory and resources, selfishness, financial greed, and the challenges of modern technology such as the misuse of social media, robotics, and artificial intelligence. There are some simple things we can do, for example to find ways to streamline work tasks, involve more people, and stop using plastics. We'll come back to how we might do more, later in the chapter.

First, we should recall that professionals are only just beginning to appreciate the uncertainty of complex systems and very little of it has permeated the thinking of non-specialists. Many still look, understandably, for simple answers to complex questions—not least the media who want yes or no answers. We seem to have lost regard for nuances and 'don't know' answers. Indeed, the latter are seen as weaknesses. As a result, politicians find it almost impossible to change their views after one pronouncement. The resulting evasions and uncertainties make people either anxious or vexatious and susceptible to rumour, gossip, conspiracy theories and fake news as we saw in Chapter 6. They seek answers that make them feel in control and which fit their worldview. These are the reasons behind how unscrupulous politicians exploit our prejudices, sometimes with messages of animosity and hatred to get elected.

One example of a particularly tricky issue is the safety of people in public health and transport, and military defence. One popular conspiracy theory is that Covid-19 was deliberately created in a laboratory. The truth is less clear and hence less palatable. How do agencies make decisions that require answers to questions like 'How do we put an economic value on human life?' 'How do we cost out losses of biodiversity and even the possible extinction of a whole species'. These are difficult and complex issues that are not easily resolved and the uncertainty surrounding them exacerbates. As mentioned earlier one currently fashionable way for economists is cost-benefit analysis (CBA) and whilst superficially attractive it is woefully inadequate.

Cost-benefit Analysis

Common sense CBA just means that when you have alternative decisions to make, for example buying a car, then you just list the pros and cons to clarify in your mind what each option offers. In recent years economists have used mathematically formalised cost-benefit analysis to quantify the pros and cons of a project in monetary terms[7]. Governments have tried to use CBA to do what the marketplace does for business. In other words, to measure, and compare in terms of money, the discounted streams of future benefits and future costs associated with a proposed project. The CBA process has several stages. In simple terms there are nine: defining objectives, listing alternatives and stakeholders, deciding what can be measured as costs and benefits, predicting outcomes, converting into a common currency, applying a discount, calculating a net present value, checking for sensitivities, and finally making recommendations based on the comparative future costs and benefits. But can everything be captured this way? The answer is no—simply because not everything has a price.

DOES LIFE HAVE A PRICE?

In their book '*Priceless*'[8] Frank Ackerman and Lisa Heinzerling demolish the monetary approach to CBA. They start by pointing out that economist's using this technique regard everything as having a price. That price can be discovered by careful questioning. But common sense tells the rest of us that for most there are matters of rights and principles that are beyond economic calculation. Human life, health and nature cannot be described in monetary terms—they are priceless. Some activities are not only unpriced but are fundamentally changed if they are marketed. Sex and voting are two classic examples. Doing it in exchange for cash (which is not unknown in either case) degrades the experience. Other questions then arise. Do we allow one person to hurt another for cash? For example, when a decision in favour of one person or group may result in harm or damage to another such as the compulsory purchase of house and land for a new railway line. Do we allow people to destroy, for financial reward, a natural resource such as an irreplaceable landscape or a harm that stretches out over decades and generations? Ackerman and Heinzerling give examples where formal cost-benefit analyses do more damage than good. One of the problems they point to is that the methods of CBA have an air of scientific certainty wrapped up in a language all its own. The risk is that basic human questions that lie at its heart are concealed, and non-specialists are excluded. They give many examples where economic theory gives opaque and technical reasons for authorities to do things that are rather obviously wrong. Cost-benefit analysis has in the recent past been used to promote deregulation. In 2000, George W Bush's advisor Terry Anderson offered a proposal to sell off all 628 million acres of public land in the USA. Behind that proposal, it seemed to be assumed that the government always does things badly whilst the private market is reliable and always does things well.

As we saw in Chapter 1 the notion of economic market efficiency goes back all the way to Adam Smith. The logic seems to imply that if the market always knows best why not sell the Grand Canyon to an oil company?

Richard Davies[9] has shown in an extensive study of 9 extreme situations that markets can be informally successful in the face of chronic experiences—but they can also fail totally. In impoverished cities and regions, such as Kinshasa in the African Congo and the Darien Gap in Central America, informal and illegal trading that avoids corrupt and voracious tax collectors, is rational when public officials like police and teachers demand payments—directly because the government does not pay them regularly. They are examples of the prisoner's dilemma mentioned earlier[23] with an absence of co-operation such that the economies get stuck in a negative loop—there is no natural bounce back. He concludes that the role of social capital is as a glue that binds the assets of society together. He calls for a new economy where resilience—informal and formal is at its core.

Free market environmentalism emerged out of a faith that a free market should be blended with improvements in the quality of the environment. The idea is attractive to conservatives and others. For example, the World Wildlife Federation (WWF)[10] has a program in southern Africa to reward native villagers who conserve elephants. The USA Oregon Water Trust[11] uses water markets to purchase or lease water for salmon and steelhead trout habitats. But water markets are rare and controversial. Property rights and appropriation rights are difficult and transaction costs are high.

CBA is a formal technique of apparent scientific certainty that appeals to people who want to reduce government regulation. In practice the adding up of the benefits of a public policy and comparing them to the costs turns out to be a study of *incomplete* costs vs *incomplete* benefits.

Huge efforts have gone into studies of costs per life and exposures to hazardous substances such as asbestos, benzene, and formaldehyde. In 2001, the USA Environmental Protection Agency (EPA) estimated the cost of a life as US$ 6.1 million[12]. They were looking at the benefits of removing arsenic from drinking water. The decision rested on the sensitivity of the valuation. At the level of the eventual choice of arsenic levels of 10 parts per billion the US$ cost health benefits were more or less equal. In 2016, the USA said that the economic Value of a Statistical Life (VSL) in the Department of Transportation Analyses was US$ 9.6 million. The VSL is the additional cost that individuals would be willing to cover for improvements in safety that, overall reduce the expected number of fatalities by 1. So that when an individual is willing to pay US$ 1,000 to reduce the annual risk of death by 1 in 10,000, she is said to have a VSL of $10 million. In 2008, EPA (the Environmental Protection Agency) placed the 'value of a statistical life' at US$ 6.9 million; in 2018 it would be about $9.6 million (based on $7.4 M in 2006 dollars). But it is based on their inefficacy—government cost to save lives. Unsurprisingly, other estimates have varied wildly. Indeed, the whole idea makes most people feel deeply uncomfortable. It is unacceptable to virtually all the world's religions. Nevertheless, the quantitative valuation of life is central to analyses of public policies based on CBA.

Attempting to put a monetary value on human life is nothing new. The first attempts were perhaps those of King Aethelbert of England around 600 AD. His legal code or dooms included the *wergild*—the compensation a killer had to pay the family of a murder victim. The King deemed that it was more expensive to kill a prince than a peasant. The medieval church sold indulgences or pardons for all sorts of crimes including murder. The church was willing to accept cash in lieu of penance. Martin Luther's challenge to the church over the practice of indulgences led to the Protestant Reformation in 1517. Another historical example was slavery. Slavery underpinned American economic development. Markets for the buying and selling of human beings were set up. Modern day examples are not so extreme but still exist. When bereaved families sue for compensation for the loss of a loved one the results vary with each case. The lawyers seem to ignore the fact that human life is the ultimate example of a value that is not a commodity. It does not have a price—its price is immeasurable. Another way of looking at compensation is to argue that figures such as US$ 6.1 million are not a price of a life but rather the price of a risk—a statistical life. But this raises the same ethical issues because no money can compensate a person when he/she is dead.

One of the premises of CBA is that a person is the best judge of his or her own welfare. But this assumption collapses when risks are involuntary. For example, who foots the bill for pollution—the polluter or the polluted? They will almost certainly be different people. A memo by Lawrence Summers (chief economist at the World Bank and later President of Harvard University) illustrates the wrong-headed thinking. He wrote '*I think the economic logic behind dumping a load of toxic waste in the lowest wage country is impeccable*'. A reply from Brazil's Secretary of the Environment included the words '*your thoughts are a concrete example of the unbelievable alienation, reductionist thinking, social ruthlessness and the arrogant ignorance of many conventional economists*'.[13]

When risks are reduced to numbers alone funny things happen say Ackerman and Heinzerling. Uncertainty collapses into a precise but inaccurate estimate of future hazards, inequity is covered up by a market with contextual characteristics that are priceless. Discounting the future can make sense when it is applied to things that really do have a price. But for the future of our environment, it is a ludicrous and dangerous strategy. No amount of money, discounted or not, will reverse the damage from delay that result in more melting of polar ice caps, more plastic waste pollution seeping into our food chain, more situations that can become suddenly and irreversibly worse such as the disruption of the flow of the Gulf Stream by arctic ice. The switch has been turned off and we cannot turn it back on. Conventional economic analyses do not seem to ever consider the possibility of a crisis. Discounting is part of this perspective. Policies that show any indifference to saving human lives are utterly politically unfeasible.

Given that we must find ways of accounting for priceless assets we now need to re-examine the various forms of capital and somehow consider them on some sort of environmental balance sheet. To check our assts against our liabilities and understand the difference as some kind of new version of net worth. We'll do that by adjusting the balance sheets that we looked at in Chapter 5.

NET WORTH IS A BETTER METRIC THAN NET PROFIT

Clearly both net profit and net worth are important for financial health. As a reminder net profit is income (or revenue) minus expenditure. A good profit will attract investors. Net worth is the value of assets minus those of liabilities owed at a point in time. It is obtained from a balance sheet that shows assets and liabilities and the balancing net worth—usually at the beginning and end of a financial year.

In business the net worth is also the shareholders' equity—the total amount that would be returned to shareholders if the company fails and all debts are paid.

Both net worth and net profit can be improved by reducing overhead expenses such as insurance, fees, rent and marketing, managing your inventory—items such as property, goods in stock and building contents, reviewing the pricing of goods and services, and reducing direct costs of operation, such as labour, materials and fuels and the costs of suppliers. Negative net worth will eventually result in bankruptcy through excessive debt. That is serious because it stays on an individual's credit report (a record of credit history maintained by a credit bureau) for many years.

Whilst financial profit is important it tells us little about how a financial entity is responding to the challenges of climate change. Profit may be being obtained by exploiting planet Earth. Net worth can include our obligations to the natural world and all forms of capital, natural, human, social, political, and environmental.

My learning point number 9 is that we should look at net worth, not profit, to assess financial value of all financial entities whether individuals, companies, or governments.

RELIABILITY OR RESPONSIBILITY AND OUR OBLIGATIONS TO THE NATURAL WORLD

We humans are fallible. Individuals can make unintended errors as slips and lapses. Mistakes are errors of judgement where the wrong thing is done believing it to be right. Violations are intentional, often well meaning, where someone purposely makes a short cut or doesn't follow a recommended procedure, to get something done. Rarely violations are malicious such as sabotage. Teams can suffer from 'groupthink' where the desire for consensus inhibits critical discussion. For all these reasons and others, we cannot expect absolute guarantees of risk, safety, and reliability. But we can and do take responsibility for the decisions we make and the actions we take. In other words, we undertake a legal 'duty of care'. By law we are expected to take all reasonable steps to ensure protection from risks and harm to our safety, commensurate with our claimed level of expertise. No-one expects a junior nurse to take responsibility for brain surgery, but a brain surgeon should take responsibility for the welfare of a patient.

If we apply that idea in a more general context, we see we have a responsibility to act based on what we know and our expertise. It becomes a metric of the dependability of a theory when applied in practice in a particular context. In short, we accept a 'responsibility to act upon a hypothesis'.

I have suggested[14] that this is a fundamental concept in which we replace the unknowable 'absolute truth' by a knowable duty of care. The measure of the quality of a hypothesis is its dependability in a context founded on strong testing, critical discussion nurtured by a good dose of ingenuity. We are accepting that measures of risk are not absolute but rather they are aids in the process of managing knowledge to control risk. Responsibility and reasonable decision making are concepts already used in the law courts.

One way in which our obligations to the natural world are being developed is by the idea of developing a circular economy.

A Circular Economy

A circular economy is one where the 3 R's of Repair, Reuse and Recycle are embedded in the ways we procure and supply goods and service. It is an attempt to work within the natural world because we are part of it. The idea is gaining ground but is one that many people are unaware of. This is despite the trend towards recycling our products and their packaging.

A circular economy is based on three principles, designing out waste, using materials longer and recycling. Goods should be produced using renewable energy, non-toxic or non-polluting ingredients, and made with a very high content of recycled raw materials. Goods should be designed to last—reused, repaired, reconditioned, or upgraded. Food waste can provide nutrients for future crops or feed for animals. If a circular economy is to be successful, then all companies in a supply chain must join in. Most of us have grown up in a linear economy—the throwaway society. A society where we mine raw materials that we process into a product and after we have used it, we dispose it in landfill or the oceans.

As I see it to change the way the majority of people think about this, we have to improve our capacity to *learn how to learn.* And we have a much to learn from the natural world.

One adage, that admittedly is losing ground, but once was an accepted truism is 'If you can't measure it, you can't manage it'. Measurement is the gold standard of science—but as we have noted many times, some things are exceedingly difficult to measure. For example, imagine you are organizing a meeting. Someone asks you how will you measure success? You may reply I will count the number of people who attend. Fine says a critic but what if they turn up, take no part—just sit there like dummies and leave? Would not a better option be to have a smaller number that actively engage in conversation and debate and then formulate some conclusions. This is just a simple example of how measurement, outside of the hard physical sciences, is much trickier than the straightforward measurement of familiar physical things like the length of a line or weight of a suitcase. In the social science there must be a clear sense of purpose if we are to define a set of good metrics.

In summary the argument is this. Performance derives from processes. If we are aiming at some behaviors and consequences, then processes need to be designed to deliver them. To do that we have to be very clear about our values and, if possible, how they might be measured as 'best value' to improve performance. To achieve that one must act to fulfil a purpose with proper duty of care. To act one must take responsibility. To accept a responsibility is to be accountable, i.e., to have to justify one's actions. To account for one's actions one must have good reasons. Good reasons are based on evidence that the purpose will be fulfilled. To evaluate evidence, one must use judgement. Judgement that is convincing is objective. To make judgement objective one must measure. For a measurement to be a justification for acting it must be dependable. Dependability is not a substitute truth but rather is a currency through which we can demonstrate the degree to which we have evidence that an action will fulfil a purpose.

Voting is perhaps the most important method for measuring the judgements of many people. Voting is the basis for democracy. But it has problems—there are many ways of organising a ballot and there are basic limitations.

BETTER METRICS ARE NEEDED

Kenneth Arrow 'Impossibility Theorem'[15] showed formally that there is no possible way in which a voting process can be rational and reasonably democratic at the same time. He seemed to have devised a theorem that undermines the whole basis of a voting democracy—until we look more closely. The democratic values Arrow considered are (a) that preferences are unrestricted and rational, (b) preferences are transitive so that if you prefer apples to pears and pears to grapes then you must prefer apples to grapes, (c) criteria are unanimous so that if every voter prefers apples to pears then the group also does, (d) there is no dictator—no one voter can force the way the others vote, (e) there is independence from irrelevant alternatives, in other words a preference for one item must not depend on other items. The standard first past the post voting system used in UK general elections does not meet these criteria as it isn't necessarily transitive. Proportional representation fares little better. Michael Scott and Erik Antonsson[16] however report that Arrow's theorem has only indirect application where choices depend on an aggregation of preferences.

Arrow also did not consider a voting system where the level of intensity of a preference is used. For example, if voters are asked how much they prefer one option over another on a scale 1 to 5. Assigning a vote of preference at 1 would mean a marginal preference whereas a 5 would mean a strong preference.

Voting is used in all democratic countries. Different methods are used but importantly the vast majority accept the methodology for their particular situation. In a moment I will suggest voting as a way of allocating obligations.

The UK Office for National Statistics (ONS) is the UK's way of providing reliable national data. It is constantly improving and developing new economic indicators such as GDP[17]. As we said in Chapter 1 and Chapter 6, GDP does not capture the living standards of people. The ONS are therefore developing new estimates of the value of human capital—the stock of skills, knowledge, and experience of the UK workforce. It is perhaps a given that higher levels of human capital can help improve wages, health, lower crime rates, and increase trust and social participation. The value of unpaid work, such as childcare, cleaning, laundry, and volunteering, is neglected. They are not included in GDP but rather in a Household Satellite Account[18]. Based on data from the national census the ONS have estimated the total value of unpaid household service work at over £1 trillion, equivalent to 56.1% of GDP. The household satellite account is vital if we are to better understanding the economy. Activities such as organising a holiday or finding the cheapest car insurance, which may have previously been carried out by the market through travel agents and insurance brokers, are now being done by individuals, using online services that are often free at the point of use. This could lead to a lower GDP but a higher sense of well-being. Various indicators are being developed to provide a more rounded and comprehensive basis for assessing changes in living standards. These include household income, spending and wealth and broader economic indicators that may also affect well-being such as unemployment and inflation. It also includes people's perceived personal financial situation—perceptions shape how well off we consider themselves to be and influences behavior. Also included are so-called 'imputed rentals'—effectively the value homeowners would get if they rented out their own homes, which is included for international comparability—as well as pension contributions, which households may not have access to for many years.

Better measures of household income and saving are needed. The ONS is using four survey questions to measure personal well-being. People are asked to respond to the questions on a scale from 0 to 10 where 0 is "not at all" and 10 is "completely". The questions are: "Overall, how satisfied are you with your life nowadays?" "Overall, to what extent do you feel the things you do in your life are worthwhile?" "Overall, how happy did you feel yesterday?" "Overall, how anxious did you feel yesterday?

Our Obligations

I have developed a theory of obligations[19] as part of principled capitalism. Tom Settle[20] first suggested the latter to replace current liberal capitalism. The Caux Round Table for Moral Capitalism and Steven Pearlstein[20] call it moral capitalism. The Caux Round Table suggests that people thrive when they live with full human dignity. Integrating worldly goods with spiritual aspirations is the foundation for human well-being and happiness. We should be striving for a moral community that provides scope and vision for our aspirations and talents. Principled capitalism is founded on the idea that moral principles should be driving our individual and collective decisions rather than utility and profit. Later I will pull together what I consider to be the ten pillars of principled capitalism (Figure 7.1).

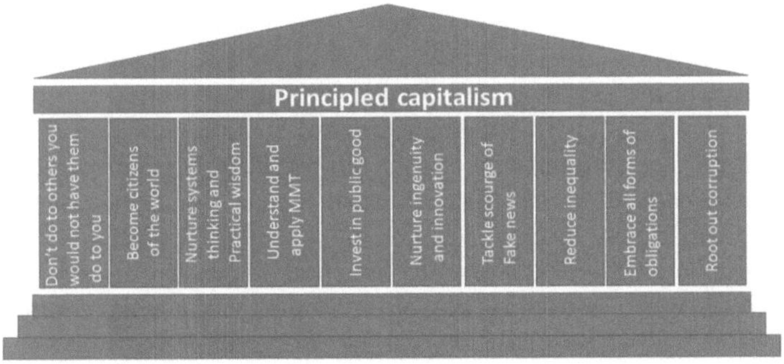

Figure 7.1 The Ten Pillars of Principled Capitalism

I have suggested a measure, obtained by voting, based on a unit called an *oblig*.[21] This is a measure of the amount of the substance of the '*stuff*' that moves around in a social field of obligation forces. In Chapter 3, I distinguished between the financial exchange worth (FEW) of part of the assets of one financial entity (the buyer/investor) for something desired to be supplied by another entity (the supplier). The objective of each entity is to increase its net worth (W) by generating a net profit (P) of income over expenditure through time. You will recall that FEW is the relative state of being useful or important enough to be desired, valued in a manner that is trustworthy and considered dependable and reliable by all entities in the system. For the most part FEW is a form of money as currency (cash and non-cash), such as the £ or $, but in theory, it could be any form of trustworthy barter. FEW is the '*stuff*' (m) of the financial system and, I suggest, is equivalent to mass or electric charge in a physical field.

The worth of obligation (TWO) is different. I defined it so that TWO entails FEW. In Chapter 3, I said that although TWO is exchangeable, it should not be tradeable in the same way as FEW and there should be no market for TWO that is not FEW. That is because TWO is a 'bigger idea' than FEW with an importantly different moral stance from that of economic utility. Whereas utility is the usefulness and degree of satisfaction of a consumer's choice, TWO is centred on a set of obligations that a financial entity has to the natural world and to those aspects of our human and social world that are beyond FEW. The idea that utility can be the basis of moral choice[22] is flawed since it assumes that ethical choice depends on usefulness to an individual rather than a duty of care for the natural world and others. However, TWO obligations are not necessarily reciprocal. Although we human beings have obligations to nature—it has no obligation to us. Nature will do what it does regardless. It takes no account of human sensibilities, but we need to treat it carefully because we are part of it. We take for granted the air we breathe and the water we drink—they are too important to be valued as FEW. Of course, there are ecosystems services to which we apply FEW such as utility companies that collect, store, clean and distribute water to our taps. The water collected by the utility companies is free but delivering potable water to our taps is not.

The worldwide scandals of poverty, inequality and social deprivation are too profound and all-encompassing to be valued as FEW. Indeed, they derive from an improper use of FEW that we need to correct. The natural capital we take for granted but which is essential for life on earth has to be measured by some other metric. We need a metric that does not depend on trade-offs susceptible to the tragedy of the commons or the prisoner's dilemma[23]. The new unit of TWO must capture the degree or amount of worth of the moral obligations we owe to each other and to the natural world. I propose that metric should be the new unit of *obligs*. An *oblig* is a way of capturing the amount of the substance of the '*stuff*' that moves around in a social field of obligation forces. In other words, as we bargain and trade with each other and our environment, we make commitments to exchange that which we consider, at the time, to be equivalents. When we trade by only exchanging FEW, then TWO *obligs* become de facto units of currency as $ or £ and their reciprocal legitimate benefits as goods and services, as a special case (Figure 3.1). In effect, TWO connects value (as financial worth) with values (as that which we consider important in the way we live) as Carney[24] advocates.

Financial worth is measured in currency (UK£, US$) and values are TWO measured in *obligs*. TWO takes us away from inward-looking selfish individual usefulness to outward-looking collaborative obligation that entails a much wider set of values than finance.

The theories of obligations and utility are similar in that they are attempts to capture the way people decide what economic and financial actions to take. But they are based on totally different values. As we saw in Chapter 1 utility captures the amount of satisfaction received from consuming a good or service whereas obligation is the giving of a binding promise, contract, or sense of duty to act in good faith in accordance with our values of legitimate standards of trust honesty as upheld by the law. Economic theories based on rational choice usually assume that consumers will strive to maximize their utility. Utility is not intrinsic quality, rather it is essentially an appeal to individual and inward-looking selfishness. By contrast an obligation is a degree of commitment to treat others as you would have them treat you. It is essentially an appeal to group relationships and outward-looking sociability. The unit of a *util* is the degree of satisfaction that a specific good or service generates for a subset of people in various situations or that a chosen course of action affords as judged by the chooser. It is contextual, for example, water is essential for life but not when someone is drowning, and an ice cream will have a larger number of *utils* in hot weather than in cold. Ordinal *utils* simply express rank choices by preferences—person A prefers X to Y, for example, A prefers to eat a pizza rather than a burger. Cardinal *utils* are measures over a set of choices, for example, A allocates 10 *utils* to eating pizzas and only 4 *utils* to eating a burger. Of course, in practice preferences may change and people may be (often are) irrational as we saw in Chapter 3. Utility theory is theoretically well developed. Von Neumann and Morgenstern used game theory to develop a theory of the maximum expected utility as long ago as 1953[25]. Faced with a set of choices an individual agent identifies the best and worst outcomes and allocates to them *utils* of 1 and 0. Any outcome between the best and worst has a probability of occurrence p. The value of p is determined by offering the chooser a bet in which he is indifferent between taking that outcome for sure or an outcome with probability p of the best outcome and (1-p) of the worst. The utility of that outcome is p. The expected (average) utility of all outcomes for each choice is used to decide between choices. The choice between bets to allocate utils is a therefore a thought experiment.

Such an experiment is not appropriate for allocating *obligs* however, since as stated earlier, the values underpinning these two measures are quite different—one aimed at the selfish interests of the consumer and the other aimed at the collective responsibilities and duties of care to others.

Balance Sheets with Obligations

I suggest that the most appropriate thought experiment to determine a quantitative measure of *obligs* is a vote—rather like the ONS questions regarding wellbeing. You will recall earlier. Figure 3.1 showing both FEW and TWO on a balance sheet. How do we measure the number of *obligs* in any situation? For example,[26]

Harold lives away from his family and feels obliged to telephone his mother. How many *obligs* does he feel he has in his emotional bank account[27] with his mother and how will his telephoning change the balance? He could ring within a week. He must ring within a month, and he ought to ring tomorrow. He could decide not to ring at all. In Harold's mind, depending on when he rings, he will deposit some *obligs* in his emotional bank account with his mother. If he fails to ring at all then he will have withdrawn many *obligs*. Of course, in real life Harold would not think it worthwhile to go to the trouble of writing a balance sheet of obligations just to ring his mother—important as that mother-son relationship is.

Another and bigger example is that, the UK government[28] has obligated that the UK will deliver a ten-point plan for a Green Industrial Revolution by 2032. The ten points include advancing offshore wind from 14 GW to 40 GW by 2030 with private investment of £20 bn and the saving of 21 $MtCO_2e$ (Million tonnes carbon equivalent) on current usage of 2,500 $MtCO_2e$. Growing low carbon hydrogen by building a capacity to produce 5 GW of low carbon hydrogen with private investment of £4 bn, saving 41 $MtCO_2e$ by 2030. Delivering new nuclear power by attracting private investment of £300m. Shifting to zero emission vehicles by installing 2,500 high powered charging points, and attracting private investment of £3 bn, saving 5 $MtCO_2e$ by 2032. Providing green public transport, cycling, and walking by government investment of £5 bn, creating a national bus strategy, and saving 2 $MtCO_2e$ by 2032. Developing zero emission aviation and shipping by government investment of £35 m and saving 1 $MtCO_2e$ by 2032. Providing greener buildings by installing 600,000 heat pumps by 2028, with government investment of £1bn, and private investment of £11bn, thus saving 71 $MtCO_2e$ by 2032. Investing in carbon capture by government investment of £1bn by 2025, thus saving 40 $MtCO_2e$ by 2032. Protecting our natural environment by government investment of £5.2 bn on flood defences, designating new National Parks and 10 landscape recovery projects. Innovating green finance by government investing £1 bn, attracting £2.5 bn private investment and launching net zero innovation portfolio, and a fusion power demonstrator by 2022. Each of the ten points contain obligations which can be captured by a number of *obligs* in a mathematical relation that tracks progress year by year[29].

This kind of analysis of government planning brings a clarity to the process that may/may not be attractive to our political leaders. Part of the lack of trust in the electorate in their elected representatives derives from this lack of clarity. Many voters, perhaps feel that in a complex world where obligations are difficult to deliver a lack of clarity may be purposely used to obfuscate the actual delivery (or not) of government promises. It is at least arguable that the kind of clarity brought by the suggested use of *obligs* will serve to build trust in the emotional bank account between politicians and the people. These kinds of measurements and analyses could be used on all forms of capital whether environmental, individual social, or political.

What is apparent, given the state of the climate change genie, is that no matter how hard we try to push the genie back into the bottle where he belongs, damage has already been done and we must deal with it, now and in the future. Another key to that is resilience.

HOW TO BECOME RESILIENT

Resilience is about recovery when things go wrong. One of the ways our natural world is resilient is through diversity. Systems that are not diverse but rely on one thing become vulnerable. When we optimise a system we consider only a limited set of criteria—and we run the risk of being vulnerable to a surprise coming from something outside of those criteria. Optimization is an idea that should be dropped and replaced by resilience. Vulnerability is the opposite of resilience. It is susceptibility to damage—and the philosophy of the weakest link. A vulnerable system is not robust, i.e., not strong, healthy, hardy, and able 'to take a knock' or persist when subject to changes or perturbations and uncertain conditions. Lack of robustness through a vulnerability is especially important when dealing with high impact, low-chance surprises such as 9/11 and the fire at London's Grenfell Towers in 2019. The weakest link in a chain defines the total strength of the chain even if every other link is very strong. A tough, strong boxer with a weak chin is only as strong as his chin.

Physicists define 'resilience' as the ability of an elastic material to absorb energy. Ecologists define it as the ability of an ecosystem to return to its original state after being disturbed, while medical doctors refer to an ability to recover readily from illness, stress, depression, or adversity. A general definition of resilience is the ability to withstand or recover quickly from difficult conditions or to adjust easily to misfortune or change. In the UK government's critical infrastructure resilience programme[30], resilience is defined as the '*ability of a system or organisation to withstand and recover from adversity*'.

A related and perhaps more commonly quoted, idea is that of sustainability. This is a capacity to endure. Sustainability implies resilience. That means a system is not sustainable when it is not resilient. But if it is resilient, it may or may not be sustainable. That is because there are other factors, such as environmental management and consumption of resources that are needed for sustainability. In other words, resilience is necessary but not sufficient for sustainability, but sustainability is sufficient for resilience. Creating a sustainable world is a social challenge that requires developments across the whole of society—economic, societal, and environmental. Sustainable living can take many forms including changing our individual lifestyles to conserving natural resources to the way we live in ecovillages and smart cities, to a greater use of permaculture, green buildings, and renewable energy.

By 2050 estimates are that 70% of the world's population will live in cities[31]. The effects of natural disasters, the need for green spaces and reducing environmental impact are imperative. Current projects include replacing hard land surfaces with green spaces to take up rainfall. Singapore is set to become a 'city in a garden' with an abundance of greenery. In Milan, Italy two residential blocks have become 'vertical forests' with 800 trees, 4,500 shrubs and 15,000 plants which if planted on the ground would cover an area the size of three and a half football pitches. Melbourne Australia is aiming to make shops, businesses, services, and leisure facilities all within a 20-minute bus ride. Miniature urban forests are being planted in Japan close to and inspired by temples. NASA is monitoring rising sea levels[32] two thirds of which is being caused by melting ice. The warmer water takes up more space and adds to sea level rise. When water warms, individual molecules move around faster, expanding the volume and causing about a third of the sea level rise. Sea level changes vary across the oceans, so some regions are more affected than others. Barriers are already being built in parts of the world. They include natural infrastructure such as reefs, marshland, mangroves, and sea grass. Sea walls, beaches and sand dunes are being built, levels of transport networks such as roads and railways are being raised. Storm water pumps are being installed and sewage systems upgraded.

Howard Dryden says that "*Life on earth depends upon healthy oceans, we have 10 years to stop toxic chemical pollution, or life on earth may become impossible*"[33] He says we are too late to stop climate change with carbon mitigation. Even if we became carbon neutral tomorrow, most life in the oceans will simply dissolve over the next 25 years, along with the food supply for 2 billion people. We will have uncontrollable climate change because we will have lost the life support system for the planet that regulates the climate. The root cause is not carbon dioxide, but toxic forever chemicals that have oil like lipophilic[34] properties that get adsorbed onto plastic particles and partially combusted carbon from the burning of fossil fuels. We need to recognise that plankton are the true lungs of the planet and our life support system. More than 60% of all life the world's oceans are plankton under 1 mm in size, and their primary food is turning into plastic and toxic particles. More than 80% of the world has no wastewater treatment. In Europe, North America and other high-income countries, the wastewater is biologically treated, but there is no tertiary treatment to remove microplastics and toxic for ever chemicals. Then the sludge is dumped on farmland or landfill sites, so effectively there is no treatment, just diffuse pollution. One solution proposed by Dryden and his colleagues is to develop community-based wastewater closed loop systems to recycle and bio-desalinate water for up to 10,000 people. 1 million of these systems would eliminate 80% of the worlds pollution and provide 10% of the food supply using wastewater.

The problem is that all these proposals take time and money, and progress is bound to be slower than required. A Fujitsu Uvance research report[35] surveyed 1,000 business and public sector leaders from 15 countries including USA, UK, China, Japan, and Thailand in 2022. It concluded that organisations are enthusiastic for sustainability but that it is not being matched by real progress. They call this the sustainability gap. For example, 61% of organisations say they are making progress, but in fact only one in 10 have made major changes to create sustainable supply chains (9%), achieving net zero (2%) and making preparations for environmental emergencies (7%). Whilst 77% of organisations say that being sustainable is the right path and good for business, a massive 18% still believe that sustainability is just a passing fad.

What is for sure is that engineering ingenuity is needed with large economic investment in infrastructure and renewable energy.

Paul Grundy, an Australian civil engineer wrote of the six steps in reducing risks[36]. First, he says, know the hazards and risks. Then identify the weaknesses. Retrofit—replace, modify, and upgrade. Plug the weaknesses and create resilience against all hazards. Plan emergency response procedures and educate the community to understand and implement the procedures. Finally rehearse emergency responses regularly. But how do we motivate people to take the need for resilience seriously when everything seems so complex, and individuals feel so helpless.

Motivating for Resilience, Common Enemy–Common Purpose

As we said earlier people come together in times of war—defeating the common enemy. This genie is the biggest common enemy we humans have ever faced. It has no respect for race, creed, wealth or any other of the many ways we find to differentiate ourselves from each other. It is a common enemy for everyone on planet Earth. But it is incipient, insidious, hidden, invisible and incredibly dangerous. The question is 'Are we up to challenging it in time'?

When people are set unrealistic goals by people who have authority over them their motivation tends to be weakened. Rules and regulations designed for average conditions can cause all sorts of problems when people are working under stress—sometimes with perverse consequences.

Our successful collective history of seeing and modelling the world as linear causality as we discussed in Chapter 2 is blinding us to our new realisation of complexity. Linear causality is dispositional—a natural tendency towards a particular condition or action, or a natural and characteristic mental model, worldview (Chapter 1) or mood. Dealing with complex interdependencies requires us perhaps to give more importance to understanding the present than imagining the future. Since there is so much uncertainty in the future perhaps, we should start our journeys with a sense of direction to where we want to get but without our goals being too specific. We should look for affordances offered as we travel and be open to embrace novelty along the pathway. Affordances are features or objects of our environment that prompt specific uses of interactions. For example, no one tells us that the indentations in a bar of chocolate are the thinnest parts of something that we can break into pieces as we wish. Recognising affordances may be the way we can reduce mental angst, help people to appreciate each other's points of view and reduce political polarization.

Helmuth von Moltke (1800–1891) was a Prussian military commander. He is credited[37] with the statement '*No plan of operations reaches with any certainty beyond the first encounter with the enemy's main force*', and '*Strategy is a system of expedients; it is more than a mere scholarly discipline*'. Perhaps the chaos of war has something to tell us regarding the uncertainty and chaos of complexity and our emergence form an age of enlightenment to one of the entanglements of complexity (and quantum physics). Nevertheless, it would be a bold and foolish person who rejected planning altogether.

Flourish or Die?

Flourishing is about achieving your potential—individually and collectively despite obstacles and setbacks. The word stems from the Latin *florere* '*to bloom, blossom, and flower*'. Aristotle was the first to understand its importance. He saw that flourishing occurs when we want to do what we ought to do—take pleasure in moral action. He thought that earthly happiness is a by-product of using our individual capacities to realize our potential. For example, creating imparts feelings of achievement. Personal hobbies like gardening, drawing, craftwork, or simply just repair jobs around the home can be very satisfying. Throughout history we have built big cathedrals, mosques, churches, and temples as expressions of communal faith. Experiencing those spaces often creates feelings of high spirituality. Sports stadia like Queen Elizabeth Olympic Park in London, exhibition halls and museums like the Guggenheim Museum in Bilbao, large scale art works such as the i360 tower in Brighton opened in 2016 and the colossal Cornish Man Engine[38] are all artistic expressions that embody the conscious use of the imagination.

To flourish together we need to maintain hope, take collective control, support each other, live with a spirit of forgiveness that unites rather than divides us and develop a sense of spirituality that goes beyond conventional religion. When we truly understand ourselves as part of the natural world then we have hope for a better future. When we admit our tendency for technical triumphalism—past arrogance in thinking that we are in total control of the natural world—then we are ready to accept unforeseen and unintended consequences and unknown unknowns. When we recognize that preparation for an uncertain future is by developing communities that integrate rather than divide us for a common purpose. When we accept that the golden rule of all religions (expressed here negatively) is '*Don't do to others that you would not have them do to you*' then we have a key to spirituality. We have to use all of our considerable ingenuity to find new ways of living to learn how to be resilient when an unknown becomes a known—for example, an unforeseen consequence (an unknown) of climate change hits us (becomes known).

Practical Wisdom

As we saw in Chapter 2 in 330 BCE Aristotle wrote in his Nicomachean Ethics *"Some people who do not possess theoretical knowledge are more effective in action (especially if they are experienced) than others who do possess it"*. I suspect that sentiment rings true with many people today. It is commonplace to hear people say 'Tommy isn't academic—he's much more practical'—as if to apologise for Tommy's lack of schooling and accomplishment. Practical wisdom is an expression of the principle of ingenuity—being inventive, resourceful, and skilful through direct practical experience. The loss of Aristotle's idea of practical wisdom in modern times means that we simply do not recognise, value, or nurture it. Practice is often criticised as being ad hoc and lacking rigour and that is indeed why the Greeks thought it inferior to science.

Rigour is the strict adherence to, and enforcement of, rules to an end. Mathematical logic is the ultimate form of absolute rigour: it has one value—truth. It is top-down reasoning, i.e., theorems (at the bottom) are deduced from axioms (including rules) which are true (at the top) by definition. The result is a self-consistent body of true statements—but only if the axioms correspond to reality. Science is rigorous in an opposite way to mathematics because it is reasoning from the bottom up. Scientists construct theories (at the top) and use them to deduce propositions (at the bottom) that they can test with an experiment.

Practical rigour is more complex than either mathematical or scientific rigour and not well understood still less appreciated. It is the meeting of a need by setting clear objectives. It is complex for many sorts of reasons. For example, Newtonian physics may not be true in all circumstances, but it is dependable for most earth-bound engineering calculations and all big bridges are designed using it. Practical rigour usually involves many values—some in direct conflict, and finding ways to reach the objectives in a demonstrably dependable and justifiable way that recognises all our rational and emotional needs and desires.

Practical rigour requires diligence and duty of care that leaves no stone unturned with no sloppy or slip-shod thinking. It also requires models that make something work, i.e., achieve its purpose and objectives. Practical rigour is about creating practical solutions to meet explicit needs and to delivering a system valued in a variety of ways, not just efficiency or cost but including such criteria as aesthetics, sustainability, practicability, and resilience.

Finally, since judgements are key then the principle of learning is also key to practical rigour—always learning to improve or self-renew. Without reflecting on and learning from experience practical problem solving becomes ad hoc.

As we face the uncertain challenges of the future, we will need people with practical wisdom as well as some of the characteristics of our super-survivors who show incredible toughness and resilience after major trauma. The biggest of our challenges are often called grand challenges. They are the barriers that if removed would lead to important advances.

GRAND CHALLENGES

In 2000 the UN established eight sweeping and ambitious Millennium Goals[39] ranging from eradicating extreme poverty and hunger to ensuring environmental sustainability. In 2015 these have morphed into 17 Sustainable Development Goals (SDG) and 169 targets. They include removing poverty and hunger, promoting good health and wellbeing, providing good quality education, having gender equality, clean water and sanitation, affordable clean energy, decent work and economic growth, industrial innovation, and infrastructure, reducing inequalities, growing sustainable cities, with responsible consumption and production, taking action on climate, life below water, life on land, and peace and justice. Finally creating partnerships to attain these goals.

Clearly such aspirational challenges require global partnerships involving everyone—again something easy to say but incredibly difficult to deliver. One of the first engineering organisations to ponder on more specific, but still high level, engineering grand challenges was the USA National Academy of Engineering[40]. In 2008 it issued its list of 14 grand challenges which essentially are a more detailed expansion of some of the 17 UN SDGs.

The first in the NAE list was to make solar energy economical. In 2023 onshore wind and solar PV power are now, frequently, less expensive than any fossil-fuel option, without financial assistance. Second was getting energy from nuclear fusion—the process that powers the sun and the stars. Fusion is the reaction in which two atoms of hydrogen combine, or fuse, to form an atom of helium and some of the mass of the hydrogen is converted into energy. There is still a lot of work to do to make it viable.

Third is to sequester carbon—to capture the carbon dioxide produced by burning fossil fuels and storing it safely away from the atmosphere. Again, there has been some progress and carbon capture is included in many climate models. The UK government supports the development of carbon capture, usage, and storage (CCUS) and has set up a task force to advise it. There is still much more investment needed.

Fourth is to manage the nitrogen cycle—processes that are central to producing food, controlling the impact of agriculture and sustainable development. Nitrogen is the main element in our air and makes up nearly four-fifths of the atmosphere. We have altered the cycle in which nitrogen in the air is '*fixed*', i.e., is converted from nitrogen gas into a form that can be used by living organisms as a nutrient. The widespread use of fertilisers and high-temperature industrial combustion has vastly increased the rate at which nitrogen is removed from the air. The resulting nitrogen oxides have caused air pollution and acid rain, polluted drinking water, eutrophication of water courses (too many nutrients, decaying algae, and depleting oxygen) and it has even worsened global warming.

Fifth is to provide access to clean water for everyone across the world. It is a scandalous lack that is responsible for so many deaths in the world—some say more than war. Estimates are that about 1 out of every 6 people do not have adequate drinking water and in some countries that proportion rises to a half of the population. Many more lack basic sanitation. The UN reports in 2022[41] that a fourfold increase in the rate of progress is required.

Sixth is to restore and improve our infrastructure. It is obvious to all that our built environment is critically important. Every day we use buildings, transport, communication, water and waste, energy, schools, and hospitals. We rely on electricity supply, delivery of gas and oil and renewable energy, telecommunications including mobile and wireless networks, railways and roads, buildings, emergency, and community. Too often we let infrastructure wear out before replacing it and simply exacerbate the problems.

Seventh is to advance health informatics and eighth challenge is to engineer better medicines and better methods of delivery of drugs. Addressing the issues of the safety of genetic modification and the problems supporting an aging population.

The ninth challenge is to reverse engineer the brain. Reverse engineering is the study of something to learn how it works in order to produce a copy, or an improved version. You are probably aware that computers have been programmed to play chess rather well. But that is a long way from being creatively human. General-purpose artificial intelligence (AI) is elusive although the media hype is considerable. Driverless cars, drones, surgical and manufacturing robots are daily news. How will we be able to certify these systems as safe? Artificial brains have been designed without much attention to real ones in a manner rather like aeronautical engineers approached flying without a great deal of learning from the flight of birds.

Tenth challenge is to prevent nuclear terror. We cannot assume that terrorists are not attempting to make nuclear weapons. Eleventh is to secure cyberspace. You would have to be a modern hermit not to be aware of the issues of personal privacy and national security in using social media through the electronic web of information sharing called cyberspace. The issues range from protecting confidentiality and integrity to deterring identity theft. It is a challenge riddled with unintended consequences such as suicides and '*sexting*' teenagers.

Twelfth challenge was to enhance virtual reality (VR). This is a computer simulation—but one for all of the senses. Thirteenth is to advance personalised learning. Some learners are highly self-motivated and learn by exploring knowledge on their own with perhaps minimal guidance. Others might prefer a more structured approach or being motivated by external rewards or may need step-by-step instruction. Some even will resist learning altogether and show little motivation or interest. Systems have already been designed that store instructional content, deliver it to students and facilitate interaction between instructors and learners. Fourteenth concerns the tools of scientific discovery. Knowing and doing develop hand in hand. Engineers make the tools that scientists use, and the knowledge gained helps engineers make better tools and the process spirals as we have seen in earlier chapters. The new bioengineering discipline of synthetic biology has exciting potential for the design of entirely novel biological chemicals and systems that could prove useful in applications ranging from fuels to medicines to environmental clean-up and more.

The fourteen challenges of 2008 concerned specific areas of application. As if to demonstrate how opinions even amongst experts differ and how times move on, in 2013 The National Engineering Academies of USA, UK and China got together to discuss challenges under the six headings of sustainability, health, education, enriching life, technology and growth, and resilience. Despite all this the UN SDGs predominate.

Progress Towards the SDGs

In 2022 The UN published 'The Sustainable Development Goals Report 2022[41]. It paints a sobering picture. The latest data reveals that the 2030 Agenda for Sustainable Development is in grave jeopardy. Multiple, cascading, and intersecting crises including Covid-19, climate change and conflict through corruption and war overshadow. The complex interactions between the crises impact on all of the SDGs. Spin-off crises are occurring in food and nutrition, health, education, the environment, and peace and security. To put the world on track to sustainability will require concerted action on a global scale. A preliminary section written by Liu Zhenmin, the UN Under-Secretary-General for Economic and Social Affairs is revealing. It is entitled '*A road map for survival*' and reads '*The severity and magnitude of the challenges before us demand sweeping changes on a scale not yet seen in human history. We must start by ending armed conflicts and embarking on a path of diplomacy and peace—a precondition for sustainable development. We simply cannot tolerate war and the senseless loss of precious lives and resources it entails. Second, we must adopt low-carbon, resilient and inclusive development pathways that will reduce carbon emissions, conserve natural resources, transform our food systems, create better jobs, and advance the transition to a greener, more inclusive, and just economy. The road map laid out in the SDGs is clear. Just as the impact of crises is compounded when they are linked, so are solutions. When we take action to strengthen social protection systems, improve public services and invest in clean energy, for example, we address the root causes of increasing inequality, environmental degradation, and climate change. Third, nothing short of a comprehensive transformation of the international financial and debt architecture will be required to accomplish these aims and to avoid a two-track recovery, with developing countries left behind. The stakes could not be higher. If humanity is to survive, we must survive together, leaving no one behind'.*

The overall conclusion is that whilst some progress has been made, for example in maternal and child health, in expanding renewable energy, and in planning for biodiversity, there is a need to accelerate progress in all in almost all of the 17 SDGs.

Why is Growth Important?

As noted in earlier chapters GDP is a measure of economic activity. A high GDP leads to higher average incomes and standards of living. People consume more, businesses hire more people and invest for the future. Unemployment is lower, absolute poverty levels are reduced, government borrowing is lower, public services are better, investment is higher, with more R&D and more choice. As we have said GDP doesn't measure a whole raft of things that improve our well-being. Growth is particularly important for developing economies to escape the worst levels of poverty and improve life expectancy. In the developed world, economic growth is less essential.

The nature of economic growth is important—if it leads to more pollution and congestion, then living standards can fall. The way it is distributed decides who benefits. If it is the rich, then growth will do little to overcome poverty. Growth can be very damaging to the environment. As a consequence, some people promote the idea of degrowth.

Degrowth

The idea of growth implies an expansion without limits. Financial growth is not an issue but there are limits to a growth in the use of the earth's non-renewable resources. The degrowth movement[42] argues that there is a need to reduce global consumption to achieve a socially just and ecologically sustainable world. That could imply the need to reduce GDP—an argument that would be very difficult for any democratic politician or economist to promote. However, it is possible, many argue, to decouple economic growth from the consumption of resources—growth without more environmental damage. The aim would be to sustain economic growth with reduced resource consumption and pollutant emissions. It would be possible to measure the reduced rate and impact of using resources for every unit of economic growth.

The need is to improve the efficiency of the consumption of energy and resource and reduce the corresponding environmental impact by switching to clean energies and reducing waste and pollutants. Supporters of degrowth highlight the importance of autonomy, care work, self-organisation, commons, community, open localism, work sharing, happiness, and conviviality—all part of Jon Alexander's Citizen Story[43].

Opponents of degrowth say that it is utopian. A transition will require a move away from the currently entrenched view of the importance of growth. As Stephanie Kelton has remarked[44] paying down government debt isn't a virtue. It's a leak. It's how money leaves the economy. She repeated what I quoted her as saying in Chapter 5 that one person's spending is another person's income. Every dollar taxed is a dollar I don't have; I can't spend, and businesses can't capture. Jason Hickel[45] thinks that degrowth and MMT go together. He sees three urgent tasks. First develop generous, high-quality universal public services including healthcare and education, public transportation, and affordable housing. The evidence is clear, he says that universal public services (rather than perpetual growth of GDP) are the key to a happy, healthy, flourishing society. Second, roll out renewable energy infrastructure quickly. We have not done this because we are told 'it's too expensive'. MMT tells us that every single government that controls its own currency can fund a rapid transition to renewables without even thinking twice about cost. What is more the costs of clean energy have tumbled as the technology is scaled up. With greater connectivity across continents costs can be reduced even more. Third, introduce a public job guarantee, so that anyone who wants to work can get a job doing socially useful things that communities need (including working in public services, building renewable energy infrastructure, and regenerating ecosystems), with a living wage, at 30 hours a week. This kind of approach reduces inequality, sees beyond everything being a commodity and ensures that everyone has access to meaningful, well-paid work and high-quality public services. Of course, inflation has, at the same time, to be controlled by making sure excessive spending doesn't overcome supply. Sceptics say that governments must maintain an artificial scarcity of money to ensure a steady flow of cheap labour for private firms. MMT elegantly disposes that argument and enables public abundance to dismantle the need for growth. MMT provides an opportunity for us to create a post-growth, post-capitalist economy.

Land and housing costs are significant barriers. Land, like air and water, is something we all need but it has been largely privatised. Market growth has relied on rents and mortgages. Land enclosure is a form of privatisation that is central to capitalism and economic growth. Degrowth, may argue must start by de-commodifying land commons for social and ecological well-being. Industrial agriculture must give way to less intensive and more sustainable agriculture. But it is not clear if any of those alternatives could feed the current and projected global population. Labour productivity of non-industrial agriculture is significantly lower.

LEARNING TOGETHER REQUIRES LEADERSHIP

Lord Browne, engineer, and one-time Chairman of BP has written[46] that respect, authenticity, and openness need to be embedded into the heart of business models. But change requires leadership. He writes '*People tend to treat companies as they do other human beings. If we trust someone, we are likely to forgive them when they make a mistake. But if that person has a bad history, we might not even give them a chance to explain*'. He continues '*Companies have to make friends before they need them...when firms increase their interactions with societal actors from a rare basis to often, they more than triple their chance of mutually beneficial outcomes. Top companies identify important stakeholders and work with them... using techniques to determine who to see and when...avoid outsiders being confused by multiple visits...with conflicting messages and...emissaries being played off against each other...Collaborating with others makes more sense when being the first mover might put the company at a competitive disadvantage...create gains that no single player can achieve individually...collaboration on standards and regulation provides the basis for industry wide reform*'.

As Greg Young has written[47], leaders shape the culture of an organisation—a large factor controlling performance and nurturing ingenuity. He calls for a new kind of 'transpersonal leader' for the 21st century where companies need to be nimble and agile in responding to rapid change. These leaders will embed authentic, ethical, and emotionally intelligent and ingenious behaviors into the DNA of the organisation, build strong, empathetic, collaborative relationships and develop a sustainable performance enhancing culture. Resilient business integrity has no place for corruption, collusion, and fraud. Leaders will see that diversity for its own sake is not the point. Diversity presents differences in thinking and performance that will be key to future success. Transparency and due diligence are ways of mitigating corruption risks. Companies need to move away from the sometimes-brutal competitiveness of the 20th century to the kind of collaborative behavior that Lord Browne identifies. Crucially Young quotes evidence that women are particularly suited to this kind of role although they often lack the self-confidence to put themselves forward.

Corporate Social Responsibility (CSR)

CSR is a business model in which a company is socially accountable to its stakeholders and the public. It is an attempt to identify the kinds of impacts the company has on all aspects of society including the economic, social, and environmental. It helps to retain staff and give them satisfaction in their work roles. It helps brand identity, bolsters customer trust and public respect. It has a ripple effect of positive good through fair pay and ethical treatment of employees. Environmentally it promotes alternative energy sources and sustainable materials with company-wide recycling programs and donating to and volunteering for local environment-focused organisations. Ethically it sets wages higher than the minimum. All materials are ethically sourced, and all employees receive competitive pay, comprehensive benefits and are treated with respect. It is philanthropic through sponsoring or donating.

CSR is now being replaced by Environmental, Social, and Governance (ESG). ESG is more proactive and takes the holistic[48] view that sustainability extends beyond just environmental issues. It is a framework to help stakeholders understand how an organisation is managing ES risks and opportunities. It arose out of health and safety issues, pollution reduction, and corporate philanthropy and is changing how many investment and capital allocation decisions are made.

Inevitably, some people bend or break the rules. One example is greenwashing. An organisation makes false, unsubstantiated, or outright misleading statements or claims about the sustainability. Some greenwashing is unintentional simply the result of a lack of knowledge or understanding, but often it is intentional marketing designed to mislead. CSR departments were often ring fenced. ESG is better integrated into the company, operating at the core, and reporting directly to the board. ESG examines the company's external footprint but also at how pollution or social protest may do harm to the company itself. Leaders are becoming aware that digital transparency and activism through social media might expose behaviors that could lead to reputational damage, regulatory fines, loss of employees, investors, and customers. However, as Gillian Tett[49] notes, events such as Covid-19 or the Black Lives Matter campaign '*have made it hard for corporate boards to ignore social welfare at a time when social media tools and increased digital transparency has made it easier than ever for activist groups to monitor what companies are doing*'. ESG has moved from being a narrow area of activism to risk management for corporate boards. Companies that neglect ESG face damage to reputation, loss of customers, investors, and employees. ESG is about self-defence and self-interest too. Unfortunately, the metrics for ESG are problematic and ratings vary wildly. Nevertheless, the social contract between companies and wider society is shifting.

Green bonds, mutual funds, Exchange funded funds (ETFs), and index funds are among the publicly traded financial instruments that make it easier for investors to align their decisions with their own ESG values. A country that is taking ESG seriously at a national level is New Zealand.

New Zealand's National Climate Adaptation Plan August 2022

The NZ plan[50] warns of '*risks that maladaptation across all domains due to the application of practices, processes and tools that do not account for uncertainty and change over long timeframes*'. It identifies four priorities for action: enabling better risk informed decisions, targeting the right locations, identifying a range of options for adaptation including managed retreat, and embedding climate resilience across government policy. It states that frameworks are needed for better decision making and long-term thinking. NZ needs to avoid costly or irreversible mistakes given the limitations on public finances and the limited time available for action.

Three objectives are set: reduce the vulnerability of assets, make all new infrastructure fit for purpose in a changing climate, and implement renewal programmes to improve the capacity to adapt. To reach these goals new guidance is needed to support infrastructure asset owners, to scope a new resilience standard or code for infrastructure that encourages risk reduction and resilience planning. They should be adapted and integrated into all Treasury decisions on new assets, renewals, and programme upgrades. Other actions include developing the National Energy Strategy, reviewing regulated network revenues, and embedding nature-based solutions. The plan outlines how the changing climate will likely affect temperature and rainfall patterns across different parts of the country. It points out that NZ is particularly vulnerable to rising sea levels. The resilience plan is part of a long-term climate adaptation strategy. We need one not just for NZ but for the entire world and all the people in it.

THE CITIZEN'S STORY–MOBILISING THE PEOPLE

We introduced Jon Alexander's Citizen story in Chapters 1, 4—a story of collaboration and cooperation that may seem idealistic and impractical. But we are facing an unprecedented existential situation. As I have said, we have a common enemy that knows no bounds of race, creed, or nationality. We have to put ourselves on a global 'war footing'. Unless we respond together, we are facing oblivion as we slowly squeeze the life out of our golden swan.

The hope lies in an international narrative that is slowly changing. It needs speeding up, intensifying, and deepening where it all seems too remote from everyday life. The reality of the threat must be made real to doubters.

In other words, to save our swan we have to mobilise the people. We must move from the consumer story to a much bigger story of who we are as human beings—the citizen's story. Finding our way through the complex challenges of the 21st century requires us to embrace the ideas, energy, and resources of everyone. It extends well beyond national boundaries. It recognises, explicitly, that all of us working ingeniously together, are smarter than any of us working alone. It educes international teamwork.

CHANGE REQUIRES LEADERSHIP

As Lord Browne says teamwork and change requires leadership. Perhaps our biggest problem is that since we have no global government, we have no unified global leadership. Our most powerful politicians in the world, including the United Nations, are our global leaders. But they are failing us as they squabble to maintain power. We need them to work together, to be, as Greg Young, says 'transpersonal leaders' that embed authentic, ethical, and emotionally intelligent and ingenious behaviors into the DNA of our communities. We need them to be the orators that inspire and unite us behind the common cause. To address our sheer survival, they must face the challenge of putting aside differences and bringing even the most hardened leaders of authoritarian states to the leadership table. Antonio Guterres, with his background in science and engineering is one of the few world leaders who seems to understand the urgency. Unfortunately, his power as Secretary-General of the United Nations looks to be limited. We need world orators to articulate the challenge. Orators like previous war time leaders such as Winston Churchill and Franklin D Roosevelt. We need them to explain complicated policies with clarity and precision, whilst entertaining and enlightening audiences with stories that range from the poignant to the humorous. We need eloquence, charm, and grace. We need epithets like Churchill's speech to the UK Parliament in 1940, '*We shall fight on the beaches, we shall fight on the landing grounds, we shall fight in the fields and in the streets, we shall fight in the hills; we shall never surrender'*. We need articulations like Roosevelt's '*The only thing we have to fear is fear itself*''. We need pronouncements such as David Attenborough's '*The future of humanity and indeed all life on earth depends on us*'[51].

WAR, CRIME, AND CORRUPTION ARE UNSUSTAINABLE FOR HUMAN FLOURISHING

Since international crime, corruption and war in its many forms benefits only a few and harms the rest of us—it is, by definition, unsustainable. As I write this in early 2023 Russia and Ukraine are sending drones to each other in a pointless exercise of wanton destruction. Intervening in such deep-seated differences, tackling crime and corruption requires world leadership and international investment. The citizen's story can help us to develop the tools from the 'bottom-up'. We should be making it clear to our leaders the people want to build empathetic, collaborative, and sustainable relationships across the whole world. It can be done. For example, the Good Friday or Belfast agreement of 1998 brought 30 years of chronic sectarian and violent divisions to a delicate lasting peace. It required massive efforts of world and local leadership. Everybody needs to feel important and cared for with no-one 'left behind'. At the national level this may be through parenting interventions, family interventions, wellbeing campaigns, and early childhood education. We need targeted programs that treat violence as a public health problem, with local frameworks, gun-free zones that foster a civic culture with better urban planning. We need early childhood interventions on hotspots that find a balance between repression and prevention, that reduce the drugs trade. We need fiscal and monetary policies that reduce inequalities.

To achieve such aims our global leaders have to commit to principled capitalism where moral principles drive our individual and collective decisions of obligation rather than utility and selfish profit.

CITIZENS ARE TOGETHER PEOPLE

The citizen story is one of 'together people' that recognises our interdependence and need for community. It's a way of reducing mental ill-health and polarization of beliefs. It's an active verb rather than a noun—a notion of doing something. Citizens identify things they can influence, find collaborators, and engage with them. Countries like New Zealand are beginning to show us the way. We need all our organisations to allow and encourage us to do so. The economy must move away from the ESG stressing stakeholder value rather than shareholder value. For these ideas to spread we must, as Alexander says, neither accept what we are given as the only possibility, nor 'throw our toys from the pram'[52] when we do not like what is on offer. Citizens propose—they do not simply reject. Citizens accept responsibility, and create opportunities for each other—they step up, and step in. They believe that to change something, you must create new models that make existing models obsolete.

As I have repeatedly said leadership is crucial to the citizen story. If those in positions of power act as if there is nothing wrong, we mistrust them even more. Citizen leaders begin by acknowledging uncertainty. They share questions and challenges with us rather than most current politicians who provide answers for us which often fail and create disillusion, apathy, and cynicism. Citizen leaders create opportunities for us to participate and contribute. They acknowledge and don't deny unknowns and more importantly unknown unknowns. They don't pretend to know what the future will look like, but they reassure us that we should build it by working together. The citizen's story must prevail, or the golden goose will die, and planet earth will become lifeless.

Is the climate change genie already out of the world's bottle? To mix metaphors—is the genie about to strangle the golden goose? We will only know for certain with historical hindsight—but then it may be too late. The genie is certainly emerging and is the biggest common enemy we humans have ever faced. It has no respect for race, creed, wealth or any other of the many ways we find to differentiate ourselves from each other and go to war about. It is a common enemy for everyone on planet Earth. But it is incipient, insidious, hidden, invisible and incredibly dangerous—are we up to challenging it in time? The signs that the genie is emerging are the disastrous effects and massive loss of life and resources through floods, fires, melting ice and other major changes to our world. Let's be clear the planet will survive—the Earth will continue to spin through space around the sun—what is at threat is all life on earth. What can we do about it all? Have we got the ingenuity to address the problem in the face of many other problems of world poverty, rising inequality, war, financial greed, and the technical uncertainty of complex systems that we are only just beginning to appreciate and address. We know what needs to be done and we know what each of us as individuals should do. But we need the systems in place to help us. For example, we can recycle plastics, but all governments need to ban single-use plastics first. In the UK plastic plates, trays, bowls, cutlery, balloon sticks, and certain types of polystyrene cups and food containers will be banned after October 2023. In the USA there are no federal regulations, but many states and cities have either enacted or are planning to enact new regulations. World-wide there are attempts to shift the responsibility for the clean-up of plastic pollution from consumers and governments to the companies that produce it.

As a formal techniques cost-benefit analysis requires everything to have a price. But there are matters of rights and principles that are beyond economic calculation. Human life, health and nature cannot be described in monetary terms—they are priceless. Unfortunately cost-benefit analysis has been used to promote a de-regulatory agenda under the cover of scientific objectivity. Many seem to assume that the government always does things wrong whilst the private market reliably does things right. We have shown that to be false. Is a person the best judge of his or her own welfare when risks are involuntary. Who foots the bill for the pollution we create collectively. We have an obligation to work within the natural world because we are part of it. To do that we must improve our capacity to learn how to learn—especially from the natural world where there is zero waste. On such is the idea of a circular economy—a concept that is gaining ground but one that much of the public is unaware. In a circular economy waste and pollution is designed out, products and materials are kept in use, and regenerated. Consumers should demand products from manufacturers that can be repaired and reused. We must design systems to be resilient—an ability of a system or organisation to withstand and recover from adversity and endure.

Voting is a fundamental for gaining consensus in a democracy. Kenneth Arrow showed formally that there is no possible way in which a voting process can be rational and reasonably democratic at the same time. This seems to undermine the whole basis of a voting democracy until we look more closely at his assumptions. It is more important to understand the present than to imagine an uncertain future we cannot predict. We should start journeys with a sense of direction but no specific goals looking for the affordances offered by the landscape and open to novelty. When people are working towards specific goals it destroys intrinsic motivation. Flourishing is about achieving our potential—individually and collectively despite obstacles and setbacks. The stories of engineering developments from the ploughs, bicycles, steam, and internal combustion engines have successfully evolved in complex patterns of ingenuities entirely embedded in society and culture—notwithstanding some 'dead-ends' and failures too. Unfortunately, the prevailing view has been that the physical and the human are distinct and disjoint. Corporate responsibility positively impacts its employee satisfaction and retention. It's also great for brand identity because CSR initiatives help bolster customer trust and public respect. ESG (Environmental, Social and Governance) is a set of standards measuring a business's impact on society, the environment, and how transparent and accountable it is. According to the UK CBI, two-thirds of investors take ESG factors into account. An ESG strategy can demonstrate that the company is reducing risks such as adapting manufacturing processes to meet future environmental legislation which could make the business a good bet for longer-term growth. The USA Business Roundtable has repudiated the Friedman monetary doctrine. They say that Americans deserve an economy that allows each person to succeed through hard work and creativity and to lead a life of meaning and dignity. They commit to delivering value to customers, investing in employees, and compensating them fairly with training and education that help develop new skills for a rapidly changing world. They say they will foster diversity and inclusion, dignity and respect and dealing fairly and ethically with suppliers and support communities embracing sustainable practices across our businesses. They will generate long-term value for shareholders with transparency and effective engagement stakeholders is essential. Gillian Tett[53] argues that the language that frames ESG is about self-defence and self-interest too. The metrics for ESG are problematic and ratings vary wildly. Nevertheless, the social contract between companies and wider society is shifting. Smith's Theory of Moral Sentiments argues that commerce functions best amid a shared social and moral framework—the rise of the stakeholder and embracing lateral vision. I am not a religious person, but my vision is for a world where we can renew the fundamental tenet of all religions. The Golden Rule—*'Do not do to others that you would not have them do to you'.* To the non-religious the rule is plain common sense in the long term.

The public need to be better informed and encouraged about what we can and cannot do about climate change and the grand challenges of the SDGs. Commerce works better in shared social and moral framework. We should support and nurture the circular economy and ESG by investing in, and nurturing of, ingenuity, and entrepreneurship. We can do this by developing skills through lifelong education and training that encourages and nurtures practical wisdom.

Net worth is a better indicator of financial health for well-being than simply focussing on profit. We should subordinate the notion of utility to a sense of obligation. We can increase supply side productivity by investment funded by borrowing but we should not borrow to cut taxes. Government borrowing puts money into the private sector and so debt is good as long as the costs of borrowing do not get too large, and inflation sets in. Ultimately the UN SDGs are the goals which we should all strive to move towards. There are small things we can all do such as insulate our homes, not buy single use plastics, avoid excessive packaging, hang clothes out to dry, eat locally produced foods in season, more vegetables, and less meat, only boil the amount of water you need, recycle everything you can from shoes to food waste. There are big things we can do if we work together such as educating and rewarding ingenuity, persuading others that we inhabit a finite Earth, so we have no choice but to rub together as we drink the same water and breath the same air. We can campaign to stop depleting biodiversity, rapid deforestation, reduce pollution[54], develop alternative fuels such as hydrogen to replace natural gas, use CO_2 to make aggregate for concrete from solid waste, develop better batteries, e.g. water, hydro power on small scale, manufacture and supply portable toilets for areas of the world where there is no proper sewerage system, measure obligation by democratic voting to value natural, human, social, and political capital. If and when we invest, we should make sure it is sustainable[55]. We should get some better economic metrics. If growth is not sustainable because we are destroying the environment and using up scarce natural resources our statistics should warn us. We should encourage organisations like the ONS in promoting better metrics that indicate environmental degradation or resource depletion, increasing inequality, middle-class suffering, or lower standards of living.

Time is short if we want to rescue our world, our golden goose—or perhaps more accurately keep it laying more golden eggs enabling life on earth to flourish. Soon we will be past the point of no return—perhaps we have passed it already. Only time will tell—I the meantime all we have is hope but only if we change the way we think and become citizens rather than consumers.

If we can persuade enough people of the need to move from the consumer story to the citizen story—we can enjoy the ride. Life is short. Human existence is longer maybe 5–7 million years. But it is less than 1% of Earth's history. Minute compared to the estimated 4.1 billion years since the formation of the planet earth and the 900 million years since the first simple multi–celled organisms appeared[56].

No one wants to be responsible for extinguishing life on earth. So, we must learn from the past, act in the present and anticipate a better future.

Collectively and individually, we must attempt to do many things to save the golden goose. But to finish this book I think seven stand out as imperative. They are understanding the economy through MMT, invest, ingeniously innovate, tackle the outbreak of fake news, reduce inequality, embrace obligations, and root out corruption.

First understand and apply MMT. MMT enables us to see that deficits and debt should not hold us back from selective investment in the future. As Richard Davies points out[9], we know a lot about where we are heading. Urbanization is just one example. With some countries having more citizens over 50 than under it then controlled investment in engineering and technology to help care for them must not be left to the free market. The polarization of views of the role of the markets must be addressed. It is not about the extremes of state controls versus entrepreneurial hunger for profit. Life in some of the world's toughest places shows why both poles should be avoided. Unsupervised markets cannot be relied on to deal with public policies. As Davies[9] writes '*Official data paints an incomplete picture*'. But he misunderstands the role of engineering in the economy when he says, '*Engineered economies stamp this (informal social capital) out and waste these skills*'. Social capital is just as important to engineers as it is to economists. It is essential in creating resilience. An economics of resilience starts with informal trade built on human and social capital.

Second, invest in the public good. The Poverty and Shared Prosperity Report 2022 from the World Bank[57] says fiscal policy is key. Policies that '*invest early in a child's life can be transformative*'. Policies that invest in infrastructure, R&D, can bring revitalized growth. Spending that addresses market failures rather subsidies is often of higher value. For example, subsidizing fertilizers, under-pricing surface, and groundwater may increase agricultural production in the short run. But in the long run they distort incentives and reduce investments in productivity. These kinds of policies can be hard to prioritize precisely because their benefits accrue over the long term and do not align with immediate political realities. They may also have less value to some households that are more present biased or governments that face high costs of borrowing. As a result, fiscal decision-making can often overlook long-run investment and growth. So first the people must persuade politicians to change the way they understand our economy through the commonsense and practical application of modern monetary theory. **We should elect leaders with climate change vision who can articulate the challenge and motivate the people.** We need climate change leaders who are eloquent orators.

Governments must invest heavily in tackling climate change through renewable energy, developing the hydrogen economy, R&D, and reskilling the population to nurture ingenuity and innovation[58]. We must persuade them to decouple financial GDP growth from the provision of universal public services that are the key to a happy, healthy, flourishing society.

And there is one further potentially powerful way of improving productivity. It is to go beyond formal education and training to help and encourage people to 'learn how to learn'. We are living in a new age of uncertainty and complexity with constant flux. The ever-increasing pace of change is demanding on all of us and many people flounder—some so seriously that they refuse to engage with public policy and or some of the latest developments in, say, IT. But herein is an opportunity. Companies and individuals that learn how to learn faster than their competitors get a considerable competitive advantage.

The key to self-learning is a 'systems thinking' approach to life that we first came across in Chapters 1 and 2. Systems thinking can be hard to grasp at first. In simple terms it is 'joined-up' thinking but it is more than that. It is an approach to understanding the world and making decisions individually and collaboratively in teams. Systems thinkers look for interconnections—they are social joiners. Out of it emerges a different approach to the processes of solving problems. Rather than isolating and fixing a problem they see the wider context and the ways in which all of the parts to the system interact. These ideas lead to new concepts such as seeing the natural and social world in layers or levels, of parts and wholes and how characteristics in those levels emerge from interactions between parts at lower levels. Another way of understanding is to see systems thinking as about getting the right information (what) to the right people (who) in the right way (how) in the right form (where) at the right time (when) for the right reason (why). If adopted by individuals, organisations, companies, and government then widespread use of systems thinking could improve efficiencies as well as providing better working conditions, more innovation and self-satisfaction—in short human flourishing.

Collectively we must try to persuade governments around the world to invest, at this time of financial crisis, in this novel form of raising productivity. We want to see the wholesale adoption of systems thinking to learn how to learn how to adapt, innovate and become ingenious in the sense that it is the root of the word engineer. It nurtures practical wisdom—being inventive, resourceful, and skillful through direct practical experience. Such moves would be a natural partner to attempts to move away from the consumer story of economics to the Citizen's story of economics as promoted by Jon Alexander and others.

Third, nurture ingenuity and innovation. Jon Alexander's Citizen story is one of collaboration and cooperation to do just this. It will need leaders, local and global, who embed authentic, ethical, and emotionally intelligent and ingenious behaviors into the DNA of communities and organisations. Those leaders must build strong, empathetic, collaborative relationships and develop a sustainable performance enhancing culture. At all levels we must recognise that all of us working together more ingeniously, are smarter than any of us working alone. And that idea applies to nation states as well. We need more than ever to nurture practical wisdom—Aristotle's idea of *phronesis* (Chapter 2) which has been lost in the mists of time. We should be moving from the consumer story to a much bigger story of who we are as human beings, embracing the ideas, energy, and resources of everyone beyond national boundaries. It requires a response to a common enemy that, as we have repeatedly stressed, knows no bounds of race, creed, or nationality. The citizen story is one of together people that recognises our interdependence and need for community. It's a way of reducing mental ill-health and polarization of beliefs. It's an active verb rather than a noun—a notion of doing something. Citizens identify things they can influence, find collaborators, and engage with them. Citizens volunteer both to help themselves and to help others.

Fourth, tackle the scourge of fake news and social media unfounded gossip. Our global leaders must make this one of their priorities since it is threatening democratic government[59]. In the USA, many years of gradually worsening economic insecurity has created deep anger. The white middle and working classes have embraced a politician with many flaws, because he challenges the present state of affairs. A leader who uses division and misinformation to promote his own goals can persuade loyalists to drift away from democratic standards—and from fact-based reality. Leaders must tackle these divisions by reducing community tensions, getting people to look out for each other, nurturing our moral imperatives and lifting horizons to see the bigger picture of the climate change challenge. They should be funding programs that help people to use social media for the general good. Programs that educate people to spot scams, be skeptical of what they read or watch, test information to distinguish it from disinformation and, perhaps most importantly, be more empathetic to those with views that differ from our own.

Rare is a person who has not received a scam telephone call, text message or email. Many elderly people are often confused and worried about some of the messages they get. They don't know how to react, and some have lost significant amounts of money. Testability is the key notion. Fake news spreads when it is not sufficiently tested. Testing does not imply that all contradictions and tradeoffs will have been eliminated but it does mean that it is the best we can do in any circumstances. It rejects harmful rational ignorance. The best way to deal with a scam is to avoid any commitment to anything suspicious (such as an unrecognized number, a request for personal information, a threat or warning, a problem with an account or a prize claim) and get advice from those whom you know you can trust.

Ultimately, the best way to test information is to measure that which can be measured and recall the four conditions of a measurement process[61]. That is what classical scientist's do whilst remembering that sometimes the measurement process interferes with that which is being measured. If measurement is not possible, then the next best thing is to subject the information to critical discussion by as many people, whom you trust, as possible. We must always remember that measured data is incomplete and only a means to an end. That end is to make the best decisions we can in a given context. There will inevitably differences of interpretation, and opinions that emerge about incomplete and complex matters but wherever possible we will know it has either been tested in the physical world or in the social world of shared knowledge. As we have said we have an important 'duty of care' to listen empathetically and respect the views of others. That is why we need principled capitalism—a kinder, fairer way of living with our obligations to each other and the natural world (Figure 7.1).

Fifth reduce inequality. Inequality undermines fairness and it gives the wealthy an unacceptable degree of control over the lives of others. Governments must use the tax system and other measures to reduce inequality wherever it exists. Taxes need to be progressive—increase as taxable income increases. Trade policies need to be open, transparent, and democratic and not decided behind 'closed doors'. One measure would be to introduce a government backed jobs guaranteed scheme as advocated by MMT supporters[60]. Illicit outflows of cash from developing countries that should have been spent on health, education, sanitation, and infrastructure increase inequalities.

Sixth embrace our obligations. We have to abandon the economics of consumer greed and grasp the mutual obligations inherent in the citizen's story (Chapter 3). An obligation is a binding promise, contract or sense of duty that should be part of economic theory. It should replace utility theory. You will recall that I referred to an obligation as an underived unit of 'stuff' in a field of human interactions analogous to mass and charge in fields of gravitational and electromagnetic forces. I distinguished between FEW as Financial Exchange Worth and TWO as The Worth of Obligation. TWO entails FEW because whilst usefulness or utility is important our obligations to each other and to our natural world are more so. Governments should require all balance sheets to list their obligation levels—possible in terms of my proposed units of *obligs* or something similar.

Seventh root out corruption and prevent wars. To reduce and eliminate corruption we have to change social norms into ones that strive to reinforce the good and reduce the bad. That means creating a greater awareness of moral disengagement and why some people act inhumanely or corruptly without being stressed. Again, this will require global cooperation and leadership. We must elect leaders who will banish corruption and unfairness by those in authority—and speak out against wars. In some countries corruption can be confused with patronage and kinship. There seems to be no international definition[62]. In its narrowest form it is bribery. In its widest form it includes extortion, cartels, embezzlement, and money laundering. Corruption is a criminal offence, involves deception, results in illegal profits, with great harm to the rest of us. In 2018 Antonio Guterres, Secretary-General of the UN quoted[63]. The World Economic

Forum estimates of the global cost of corruption as, at least, $2.6 trillion, or 5 per cent of the global gross domestic product (GDP). He added that the World Bank estimates that businesses and individuals pay more than $1 trillion in bribes every year. "*Corruption robs schools, hospitals and others of vitally needed funds*", he said, adding that it also rots institutions—even as public officials enrich themselves or turn a blind eye to criminality—while also depriving people of their rights, driving away foreign investment, and despoiling the economy. He continued, "*Corruption breeds disillusion with Government and governance and is often at the root of political dysfunction and social disunity*", noting that the poor and vulnerable suffer disproportionately from it. Corruption causes bloodshed, spreads diseases, undermines democracy.

War is a distraction in our bigger battle with climate change. It is like shifting the deck chairs on the Titanic. We have to help those who see war as a means of getting what they want as inevitably something that we, everyone on the planet, will all lose out on. We must negotiate a way out of war for the sake of the entire human race. War is discomposing and a major contributor to killing the golden eggs of the golden goose—the destruction of humanity for short term gain.

We must re-emphasize the golden rule of all religions is also commonsense for everyone. It is the basic rule for humans flourishing. *'We should not do to others that we would not have them do to us'.* Corruption and war have no place in the citizen story and the flourishing of life on earth.

My final learning point number 10 is that principled capitalism (Figure 7.1) can emerge out of liberal capitalism if we widen utility as personal satisfaction and the profit motive to include all forms of obligations to each other and to nature. Through it we can become citizens of the world. We can be defined by our interdependence and community.

We should not cower from those who say that these aims are naïve, idealistic, and impractical. We work with an understanding that everything exists in relationship to everything else—nothing exists in isolation. In the face of a common enemy, we must all recognize the need for, and nurturing of, practical wisdom and be ingenious social, political, environmental, and physical economists and engineers.

End Notes

1. John F. Kennedy is quoted as saying this at https://www.brainyquote.com/quotes/john_f_kennedy_103820
2. Joseph Stiglitz: See https://www.theguardian.com/commentisfree/2019/nov/24/metrics-gdp-economic-performance-social-progress
3. See report at https://polarbearscience.com/2019/02/09/polar-bears-have-been-terrorizing-a-russian-town-on-the-barents-sea-since-december/
4. Hawken, P. 2021. Regeneration, Penguin. See https://regeneration.org/the-book
5. Holmes-Gentle, I., Tembhurne, A., Suter, C. and Haussener, S. 2023. Kilowatt-scale solar hydrogen production system using a concentrated integrated photoelectrochemical device. Nature Energy. See https://www.nature.com/articles/s41560-023-01247-2

6. Australian Government, Dept of Climate Change, Energy, the Environment and Water 2022. Powering Australia. See https://www.energy.gov.au/government-priorities/australias-energy-strategies-and-frameworks/powering-australia
7. UK Government, Department for Work and Pensions, Research and Analysis. Cost benefit analysis, technical report at https://www.gov.uk/government/publications/an-evaluation-of-the-group-work-jobs-ii-trial/cost-benefit-analysis-technical-report#:~:text=The%20CBA%20examines%20all%20the,an%20assessment%20of%20the%20benefits.
8. Ackerman, F. and Heinzerling, L. 2004. Priceless. The New Press, New York.
9. Davies, R. 2021. Extreme Economics. Black Swan, Penguin, UK.
10. World Wildlife Federation (WWF) has a program in southern Africa to reward native villagers who conserve elephants. See https://www.worldwildlife.org/species/african-elephant
11. The USA Oregon Water Trust uses water markets to purchase or lease water for salmon. See https://water.oregonstate.edu/sites/water.oregonstate.edu/files/oregon_water_markets_v3_0.pdf
12. According to Ackerman and Heinzerling the USA Environmental Protection Agency (EPA) estimated the cost of a life as US$ 6.1 million when they were looking at the benefits of removing arsenic from drinking water. For a more recent figures see https://www.epa.gov/environmental-economics/mortality-risk-valuation#whatisvsl
13. See http://www.whirledbank.org/ourwords/summers.html
14. I have suggested in D.I. Blockley. 1985. Reliability or Responsibility. J. Structural Safety. 2(4): 273–280, that this is a fundamental concept in which we replace the unknowable ‘absolute truth’ by a knowable duty of care. See also Blockley, D.I. 1980. The Nature of Structural Design and Safety. Ellis Horwood, Chichester, UK.
15. Arrow’s Theorem (Stanford Encyclopaedia of Philosophy) showed formally that there is no possible way in which a voting process can be rational and reasonably democratic at the same time.
16. Scott, M.J. and E.K. Antonsson. 1999. Arrow’s Theorem and Engineering Design Decision Making. Research in Engineering Design. 11: 218–228. https://doi.org/10.1007/s001630050016. They report that Arrow’s theorem has only indirect application where choices depend on an aggregation of preferences.
17. The UK Office for National Statistics (ONS). See https://blog.ons.gov.uk/2018/10/01/beyond-gdp-how-ons-is-developing-wider-measures-of-the-uk-economy/.
18. For the Household Satellite Account. See https://www.ons.gov.uk/economy/nationalaccounts/satelliteaccounts/articles/householdsatelliteaccounts/latest.
19. I have developed a theory of obligations as part of *principled capitalism*. See Blockley, D.I. 2022. Exchanging Obligations: Accounting for All Forms of Capital. Journal of Interdisciplinary Economics. 1–22. https://10.1177/02601079221081717
20. See also Chapter 3. Principled capitalism is founded on the idea that moral principles should be driving our individual and collective decisions rather than utility and profit. See Settle, T. 1976. In search of a third way. McClelland and Stewart, USA. Settle, T.W. 1969. Scientists: priests of pseudo-certainty or prophets of enquiry? Science Forum. 2(3): 21–24. The Caux Round Table for Moral Capitalism is at https://www.cauxroundtable.org/case/. They propose seven core principles for responsible business as respect all stakeholders, contribute to economic and social development, build trust, respect rules and conventions, support responsible globalization, respect the environment, and avoid illicit activities. Their eight principles for government

include building trust, having discourse, maintaining civic order, with no corruption, good security and liberty, justice, welfare, with global cooperation. They adduce ten principles for good citizenship and six for ownership of wealth including supporting social capital, investing with no abuse of power. They advance eight principles for non-government organisations including maintaining integrity, working for public benefit, being transparent with good governance, independence, respect for the law, truthfulness, and accountability. Donald Delves 2012. Principled Capitalism: A Pragmatic Approach to Corporate Governance Principled Capitalism, proposes that capitalism should operate using four core principles: ingenuity and entrepreneurialism., fair rewards to all, responsibility for impact, and stewardship of businesses, communities, and the environment. See https://www.forbes.com/sites/donalddelves/2012/10/10/principled-capitalism-a-pragmatic-approach-to-corporate-governance/?sh=57ae2e3542d2. Steven Pearlstein says we need to find a better balance between conflicting moral obligations. He calls for limiting special interest money in politics, bolstering low incomes, national service such as AmeriCorps, reduced concentration of power and money, and equalising educational opportunities across the USA. Pearlstein, S. (2018). Moral Capitalism; Why Fairness Won't Make Us Poor. St Martin's Press, New York.

21. I have suggested a measure, obtained by voting, based on a unit called an *oblig*. See Blockley, D.I. 2022. Measures of Obligation. Cogent Business & Management. 9(1). See https://www.tandfonline.com/doi/full/10.1080/23311975.2022.2154101#:~:text=The%20theory%20of%20obligations%20in%20principled%20capitalism%20%28Blockley%2C,a%20social%20field%20att%20of%20dynamic%20obligation%20forces.
22. Hirsh, J.B., Lu, J.G. and A.D. Galinsky. 2018. Moral utility theory: Understanding the motivation to behave (un)ethically. Research in Organisational Behavior. 38(1). See http://dx.doi.org/10.1016/j.riob.2018.10.002
23. For an outline of the tragedy of the commons and the prisoner's dilemma see King, M. 2020. The End of Alchemy & Radical Uncertainty Decision-making for an unknowable future. Abacus/The Bridge Street Press, UK.
24. Carney, M. 2021. Value(s). William Collins, UK.
25. Neumann, J. von and O. Morgenstern. 1953. Theory of Games and Economic Behavior. Princeton, NJ. Princeton University Press.
26. *Ibid.* [Blockley, End Note 18].
27. Covey, S. 1989. The 7 Habits of Highly Effective People, Simon and Schuster, UK. See also https://www.shortform.com/blog/emotional-bank-account-examples-7-habits/
28. HM Government UK 2020. The Ten Point Plan for a Green Industrial Revolution. See https://assets.publishing.service.gov.uk/government/uploads/system/uploads/attachment_data/file/936567/10_POINT_PLAN_BOOKLET.pdf
29. It is crucial to examine the assumptions and axioms of any mathematical relation. Beware the error of neglecting a crucial assumption that takes the relation out of the context of your problem.
30. Sector Resilience Plan for Critical Infrastructure. 2010. UK Cabinet Office. See https://assets.publishing.service.gov.uk/government/uploads/system/uploads/attachment_data/file/271335/sector-resilience-plan-2010.pdf. Also The UK Government Resilience Framework 2022. The Cabinet Office. See https://www.gov.uk/government/publications/the-uk-government-resilience-framework/the-uk-government-resilience-framework-html which seeks to develop and share understanding of the civil contingencies risks we face, to focus on prevention rather than cure wherever possible and put a greater emphasis on preparation and prevention. It says that resilience is a 'whole of

society' endeavour, so we must be more transparent and empower everyone to make a contribution.

31. For the 7 innovative projects making cities more sustainable. See https://www.weforum.org/agenda/2020/09/cities-sustainability-innovation-global-goals/#:~:text=1%20It%E2%80%99s%20estimated%20nearly%2070%25%20of%20the%20world%E2%80%99s,tackle%20challenges%20such%20as%20climate%20change.%20More%20items.
32. NASA is monitoring rising sea levels. See https://www.nasa.gov/specials/sea-level-rise-2020/?linkId=160910552
33. See Howard Dryden at https://www.goesfoundation.com/
34. Lipophilic is the tendency to combine or dissolve in lipids or fats.
35. Fujitsu Uvance and F.T. Longitude 2022. Organisations have a sustainability gap. See https://sustainability-transformation-ft.global.fujitsu.com/home/
36. Paul Grundy, an Australian civil engineer wrote of the six steps in reducing risks. See Grundy, P. 2011. Disaster Risk Reduction: The Engineer's Role. IEAust. Engineers, Australia, Barton, ACT, Australia.
37. See Helmuth von Moltke (1800–1891), a Prussian military commander at Oxford Reference Essential Quotations (4th ed.). Oxford University Press at https://www.oxfordreference.com/view/10.1093/acref/9780191826719.001.0001/q-oro-ed4-00007547
38. A man engine was a mechanism of reciprocating ladders and platforms for miners to travel up and down between working levels during the nineteenth century. It was invented in Germany but used in the tin and copper mines in Cornwall until the beginning of the twentieth century. The Cornish Man Engine puppet is a 10-metre-high giant mechanism built in 2016 to mark the 10th anniversary of the Cornwall and West Devon Mining Landscape becoming a World Heritage site. It toured from Tavistock to Geevor Tin Mine in July and August 2016.
39. The Sustainable Development Goals (SDGs) were agreed in 2015. They are 1. End poverty in all its forms everywhere. 2. End hunger, achieve food security and improved nutrition and promote sustainable agriculture. 3. Ensure healthy lives and promote well-being for all at all ages. 4. Ensure inclusive and equitable quality education and promote lifelong learning opportunities for all. 5. Achieve gender equality and empower all women and girls. 6. Ensure availability and sustainable management of water and sanitation for all. 7. Ensure access to affordable, reliable, sustainable, and modern energy for all. 8. Promote sustained, inclusive, and sustainable economic growth, full and productive employment, and decent work for all. 9. Build resilient infrastructure, promote inclusive and sustainable industrialisation and foster innovation. 10. Reduce inequality within and among countries. 11. Make cities and human settlements inclusive, safe, resilient, and sustainable. 12. Ensure sustainable consumption and production patterns. 13. Take urgent action to combat climate change and its impacts. 14. Conserve and sustainably use the oceans, seas, and marine resources for sustainable development. 15. Protect, restore, and promote sustainable use of terrestrial ecosystems, sustainably manage forests, combat desertification, and halt and reverse land degradation and halt biodiversity loss. 16. Promote peaceful and inclusive societies for sustainable development, provide access to justice for all and build effective, accountable, and inclusive institutions at all levels. 17. Strengthen the means of implementation and revitalise the global partnership for sustainable development.
40. National Academy of Engineering. Grand challenges for engineering. See http://www.engineeringchallenges.org/

41. UN 2022. The Sustainable Development Goals Report 2022. See https://unstats.un.org/sdgs/report/2022/The-Sustainable-Development-Goals-Report-2022.pdf
42. The degrowth movement—can we save the planet by shrinking the economy. See https://www.vox.com/future-perfect/22408556/save-planet-shrink-economy-degrowth
43. Alexander, J. 2022. Citizens: Why the Key to Fixing Everything is All of Us. Canbury, UK.
44. Stephanie Kelton April 2020. 'They're going to have massive deficits. And it's fine': See https://www.ft.com/content/ea25934a-7b28-11ea-af44-daa3def9ae03
45. Jason Hickel 2020. Degrowth and MMT. See https://www.jasonhickel.org/blog/2020/9/10/degrowth-and-mmt-a-thought-experiment
46. Browne, J., R. Nuttall and T. Stadlen. 2015. Connect, Virgin Digital, UK.
47. Greg, Y. 2022. See https://www.leadershapeglobal.com/greg-young
48. Holistic generally refers to the whole being more than the sum of the parts—for example in medicine treating the person as well as the affected parts. Here we go one stage further in thinking of systems as sets of interacting holons which are both parts and wholes, arranged in layers. See Note 15 Chapter 1.
49. Gillian, T. 2022. ESG exposed in a world of changing priorities. FT. See https://www.ft.com/content/6356cc05-93a5-4f56-9d18-85218bc8bb0c.
50. New Zealand Government .2022. Adapt and thrive: Building a climate-resilient New Zealand–New Zealand's first national adaptation plan. See https://environment.govt.nz/publications/aotearoa-new-zealands-first-national-adaptation-plan/ warns of the risks of not adapting to climate change.
51. BBC. 2020. David Attenborough: Five quotes to make you think. See https://www.bbc.co.uk/newsround/51131823
52. *Ibid* [Alexander].
53. *Ibid* [Gillian Tett].
54. As someone has already noted if CO_2 smelt like sewage people would want to reduce it much more quickly.
55. If and when we invest, we should make sure it is sustainable. See https://www.abrdn.com/en-gb/intermediary/sustainable-investing?gclid=Cj0KCQjw1vSZBhDuARIsAKZlijRfkmZ_S3a70stKoBnNeKN7jxeVgitD5zXR23Qj8BhyvvGvyNgvRmAaAijaEALw_wcB.
56. See https://humanoriginproject.com/evolution-and-timeline-of-life-on-earth/
57. The World Bank 2022. Poverty and Share Prosperity 2022: Correcting Course. Washington DC, USA. See https://www.worldbank.org/en/publication/poverty-and-shared-prosperity
58. We have already developed many of the engineering and technologies—but they are not all scalable in the required time frame. The consequent changes could be quite disruptive—just as the motor car replaced the horse, the internet has altered the way we communicate, and online shopping is changing the way we buy things. The profile of the future work force will be radically different. That is why governments must *invest now in new technical skills training*. For example, the transition to electric cars requires a range of retrained mechanics and technicians. The hydrogen economy requires technicians who can maintain the new home heating systems. Apprenticeships must become the norm for engineering technicians so that they can 'learn on the job', develop practical ingenuity and flexibility to deal with rapid adoption of digital systems. Repair should be the norm rather than the exception and made much easier by industrial

designers. Chartered engineers will need to continue to do the R&D required to find practical innovative ways to deliver the energy systems that can make energy available when required—that will require practical storage systems. A great deal of ingenuity and innovation has to be funded if the existing engineering technologies are to be delivered on time.

59. Mason, L., J. Wronski and J. Kane. 2021. Activating Animus: The Uniquely Social Roots of Trump Support. America Political Science Review. 115(4): 1508–1516. https://doi:10.1017/S0003055421000563. They write '*Partisanship in American politics is inextricably linked with social identities, and sentiments toward party-aligned groups affect political orientations. However, out-group animosity may operate differently depending on the party or elite. We investigate the extent to which citizens' animus toward (Democratically aligned) minority groups drove political support for Donald Trump, whose incendiary rhetoric regarding such groups is unique in modern presidential politics. Leveraging panel data beginning before Trump's candidacy, we find that animus toward Democratic-linked groups in 2011 predicts future support for Trump regardless of party identity. This animus does not predict future support for other Republican or Democratic politicians or either party. Nor do we find that animus toward Republican groups predicts support for Democratic elites. Trump's support is thus uniquely tied to animus toward minority groups. Our findings provide insights into the social divisions underlying American politics and the role of elite rhetoric in translating animus into political support'.*
60. See End Note 7 of Chapter 5.
61. The four conditions of a measurement process were set out in End Note 18 of Chapter 3 and in Chapter 6 sub-section Measurement.
62. McKittrick, R. 2022. Personal Note. Bob retired as a Director of Scott Wilson plc, an international civil/structural engineering and planning consultancy in 2007. He has done voluntary work for the UK Methodist Church, was President of the Institution of Structural Engineers 2002–2003. In 2004 he helped found the UK Anti-Corruption Forum, an alliance of UK business associations, professional institutions, civil society organisations and companies with interests in the domestic and international infrastructure, construction, and engineering sectors. The Forum's objective is to help create a business environment that is free from corruption. See https://www.anticorruptionforum.org.uk/
63. United Nations, 8346th Meeting. 2018. Global Cost of Corruption at Least 5 Per Cent of World Gross Domestic Product. Secretary-General Tells Security Council, Citing World Economic Forum Data. See https://press.un.org/en/2018/sc13493.doc.htm

Appendix

The Flow of Money

'Everything must come from somewhere and then go somewhere'—Wynne Godley

How is money created and where does it go? What are the transactions between financial entities such as individuals, companies, banks (domestic and foreign), governments and investors?

In Chapter 5, we saw that balance sheets are important not just as a snapshot at a point in time of a company's accounts but because they are essential to understanding the flow of money between financial entities. I referred to Godley tables as developed by Steve Keen to model stocks and flows of money. Now let's look at those stocks and flows in a little more detail[1].

STOCKS

A stock or equity is a security that is part of owning a company. Shares are units of stock. They entitle the owner to a proportion of the corporation's assets and profits. Stocks are traded on stock exchanges.

A more general interpretation of a stock is a store of something. In economic stock and flow models, stocks are stores of money in various accounts and money flows between stocks.

You will recall from Chapters 1 and 5 that money comes in different forms, M0 to M4. Cash, M0 and M1, is only one form of money. Indeed, physical currency such as banknotes and coins are only about 3% of the money in an economy. Around 80% of all money is held electronically in banks as deposits. Banks hold electronic money in an account at a central bank—called a reserve account.

Physical currency, especially paper lasts only a few years. In actual fact it's not literally paper—old notes were 75% cotton and 25% linen fibres. Modern UK notes are made of a thin plastic polymer with a see-through window and holograms. Damaged currency is taken out of circulation and destroyed. The central bank orders the printing of replacement paper notes, and the banks pay for them from their reserve accounts at the central bank.

New money is created when you borrow from a bank. The bank credits your deposit account at the bank electronically. The money didn't exist before. You can then withdraw it as cash or spend it electronically using, for example, a credit card or bank transfer as you wish. As you pay back your loan, as you must, the electronic money is deleted and hence no longer exists—it is destroyed. Banks can't lend money this way without limit. They are regulated. They are required to hold a certain amount of money as investment capital in their reserve account at a central bank—in case people default on their loans. Banks will go bust if they lend too much. Naturally, people usually borrow money to spend. Consequently, the money they borrow may end up in other banks. The banks will need to transfer that money between their respective accounts at the central bank. Therefore, if one bank lends out too much money it will not have enough money in its central bank account to pay other banks.

So, when a bank creates money, it doesn't actually create it out of thin air as it first may seem. In fact, it creates money out of assets of third parties, for example, a credit card company, a bank, and a central bank. If you pay for a coffee using a credit card, you are effectively giving an asset—your future repayment—and in return you get a credit card company IOU backed by a bank. These IOUs are accepted as a means of payment (i.e., a form money). The credit card company and bank are trusted to hold healthy assets as well as reserves in an account at a central bank. That privately issued IOU money is therefore asset-backed. It can be traded at par with central bank money as currency or reserves. So, the solvency of a bank is a constraint on the creation of money.

HOW IS MONEY EXCHANGED?

There are many different systems around the world for exchanging money between financial entities. They vary in scope and size—some are direct transfers, some are within one country, and some are international. In the UK, widely used systems are BACS (Bankers Automated Clearing System), and CHAPS (Clearing House Automated Payment System) and Faster Payments. An international system is called SWIFT (The Society for Worldwide Interbank Financial Telecommunications). Nowadays all the major banks in a country hold accounts with a central bank. They move money between themselves simply by instructing the central bank to debit one account and credit the other. When you ask your bank to transfer money to your friend Cynthia, the bank sends that instruction, for example, to Faster Payments. They forward the instruction immediately to Cynthia's bank who then credit Cynthia's account—the so-called clearing process. The next stage is called the settlement process. Faster Payments sends instructions to the central bank to debit and credit the two banks. Of course, this is just a bare outline, in reality the processes are more complex. For example, banks are required to commit dedicated collateral to ensure that no bank tries to send out more money than they have available in their central bank account.

STOCK AND FLOW MODELS ARE PROCESSES

Stock and flow in physical systems are driven within fields of forces. In a moment we will find a direct analogy between physical and financial stock and flow. Let's start with physics. A physical field is a region of space that is under the influence (i.e., experiences a force) of some physical agency such as gravity or electromagnetism. The work, or energy transferred per unit mass (m), to move an object in a gravitational field or an electric charge in an electromagnetic field is called a potential. Potential is state of the possibility of becoming (change)—it is available energy to do work. A body or system of bodies is in a state of equilibrium when the forces on it balance—but that state has the potential for change through perturbations. Stable equilibrium occurs when perturbations result in a return to the same state—unstable equilibrium occurs when they do not. The state of an engineered system is a therefore dynamic process. Engineers design arrangements of natural assets to control the process and do something useful. In a gravitational field, the potential is the velocity of a mass, and the flow is force. In an electromagnetic field the potential is voltage, and the flow is current or change of electrical charge. Flow is inhibited by storage of potential to release flow (as in spring or an electrical inductor–inductance), storage of flow to create potential (as in a flywheel or a battery–capacitance) and dissipation of energy (as in frictional damping and electrical resistance).

By analogy there are fields of forces within human interactions[2]. In Chapter 3, I said that an obligation is a binding promise, contract or sense of duty that ought to be used instead of utility. I referred to an obligation as an underived unit of 'stuff' in a field of human interactions. This 'stuff' of obligation is analogous to mass and charge in fields of gravitational and electromagnetic forces. I distinguished between FEW as Financial Exchange Worth and TWO as The Worth of Obligation. FEW is the relative state of being useful or important enough to be desired, valued in a manner that is trustworthy and considered dependable and reliable by all entities in the system—usually money. TWO entails FEW but goes beyond it by recognising that whilst usefulness or utility is important our obligations to each other and to our natural world are more fundamental.

In an economic system potential is a need or want (demand) and flow is the exchange of financial worth (goods, services, and money). Godley tables as developed by Steve Keen are a form of stock flow chart similar in layout to a balance sheet. The tables show how FEW money flows between accounts. By contrast balance sheets show financial values frozen at a point in time. Like balance sheets Godley tables have columns representing various accounts as assets, liabilities, or equity (net worth). Since assets minus liabilities equals net worth, the rows must sum to zero. The sum of the columns represents the net flow through that account. The values in each cell of a table are completed rigorously using double entry bookkeeping.

To illustrate the power of this approach we will explore a very simplified model that includes the Treasury (as the government ministry charged with maintaining control over public spending and economic policy), the private sector (mainly commercial banks), the public sector[3] and the central bank. We will consider only financial assets. Nonfinancial assets such as a house, gold, market value shares (a share is only a liability for the issuing company up to its issue price), cryptocurrencies, etc., are excluded. Of course, most people see their houses and precious metals as financial assets. But as they are illiquid and take some time for their value to be realised—they are generally classed as termed nonfinancial assets.

For every transaction money comes from somewhere (one account) and goes somewhere else (another account) and we want to trace the flows of money between the two accounts. Figure A.1 shows a much-simplified model of just some of the accounts held by these four UK major financial sectors. The Treasury (representing government), the Bank of England (a central bank), commercial banks (of which there are many) and the public sector (lots of individuals and companies). Each sector holds an account matched by an opposite account in another sector. For example, the Public deposit account (an asset) is matched by a liability account in the Banking sector. Banks reserves account (asset) corresponds to a liability reserves account at the Central Bank. Reserves are, in effect, deposit accounts of banks at the Central Bank. The Central Bank has asset bonds accounts as liability accounts at the Treasury. The Treasury account is in effect the Central Bank's deposit account at the Treasury. The Treasury facilitates government spending and initiates taxation. Taxes increase the net worth of the Treasury whilst government spending decreases it. In other words, a spending deficit of (Spend–Tax) decreases the net worth of the Treasury. Over time the Treasury's account at the Central Bank will go negative if deficits continue with an eventual need for an overdraft. This is where the issuing of Treasury Bonds (UK gilts) plays a role as we will see in a moment.

Every transaction requires four entries in a Godley table—two in one sector with a corresponding pair in another sector. For example, an increase in assets at the Treasury by receiving taxes—a credit—results in a rise in equity or net worth of government. In Figure A.1 the flows are labelled using variables. The variable Spend represents government spending and Tax is total tax receipts. Subscripts to flows show the source and superscript show the destination. For example, flows for sales of bonds by the Banking Sector (B) to the Public (P) are called Sell_{B}^{P}. Another example is Buy_{T}^{B} which is the flow of money when banks buy Bonds from the Treasury (T). Purchases of bonds by the Central Bank (CB) from the Banking Sector are $\mathrm{BuyBonds}_{B}^{CB}$. Finally purchases of bonds by the Central Bank from the Public are labelled as Buy_{P}^{CB}. Interest payments are Int_{B}^{B} and Int_{B}^{P} payable to banks and the public, respectively. Lend_{CB}^{T} is the flow of loans from the Central Bank to the Treasury.

What does Figure A.1 tell us? We have already seen that taxes increase the net worth of the Treasury whilst government spending decreases it and a spending deficit of (Spend–tax) decreases the net worth of the Treasury. But at the same time the net worth of the public increases. If you follow up why this is so in Figure A.1 you will see that the Treasury assets at the Central Bank are matched by Central Bank reserves which in turn are held by banks and reflected in public deposits. So why does the Treasury buy bonds to cope with a deficit? First it sells bonds of a value exactly equal to its deficit. It sells to certain banks as we noted in Chapter 5. In turn the banks sell to other banks, to the Central Bank and to the public. When certain banks sell bonds to Central Banks, the government deficit is unaffected. This is because the sale reduces the monetary value of the total bonds held by the banks and replaces them with the same amount of reserves. The banks hope to profit on the sale through a difference between purchase and sale price but the sale itself is simply a way of exchanging Bank bond assets for reserve assets. When banks sell bonds to the public then public deposits diminish.

So, the net result is that the Treasury gets the deficit back but at the cost of having to pay interest. Of course, deficits leading to overdrafts are a serious problem for you and me as ordinary customers of an ordinary bank. An overdraft will attract a heavy interest rate and you risk bankruptcy. Things are different for the Treasury of a government with a fiat currency. The Treasury effectively owns its Central Bank although they operate independently. The Central Bank could let the Treasury operate with an overdraft. In that case the government would not have to sell Treasury bonds at all. However, this is legally forbidden in most countries except in exceptional circumstances like the pandemic. This is why there is a 'magic money tree'. A currency-issuing country can create money by running a deficit. It does not have to borrow from either private banks or the public to finance that deficit. But legally in most countries it is required to do so.

So contrary to most people's understanding a government deficit doesn't take money from the public. Rather it actually puts money into the hands of the public. The government does not have to borrow from the public to finance its deficit but because it sells bonds exactly equal to its deficit. But of course, it has to pay interest on those bonds. How does it do that? It sells even more bonds to the Central Bank—shown as $\text{Lend}_{CB}{}^{T}$ in Figure A.1.

That lending from the Central Bank to the Treasury is commonly called the cost of government borrowing. The name is misleading since it invokes in the minds of many, the kind of overdraft that we might have as individuals. Government selling bonds is a savings mechanism that looks like a loan. It is just like you buying a bank or building society bond to save your money—you lend them money and they give you an IOU bond as a promise to pay you back in due course. The only difference is that when you buy government bonds you know that they are totally safe for the saver. As indicated earlier a better name for so-called government borrowing might be public savings or, if you prefer, deferred, or roll over borrowing. The Treasury issues IOUs and receives money. But essentially it is swapping present financial value for future value. Future value is represented by the level of interest rate paid out for the loan. Whether this is inflationary depends on the economic conditions when it is done.

Out of all of this we can draw four more learning points. The first is that a government spending deficit creates net worth for the non-bank public. The second is that interest on government bonds creates net worth for the banking sector. The third is that the deficit creates reserves for the banking sector, and those reserves are what banks use to buy government bonds. Fourth, the costs of government borrowing are deferred or rolled over borrowing which may or may not be inflationary.

Treasury

	Asset	Liability				Net worth
	Treasury	$Bonds_B$	$Bonds_P$	$Bonds_{CB}$	$Loans_{CB}^T$	$Treasury_E$
Government Spend	-Spend					-Spend
Government Tax	Tax					Tax
Sell Treasury Bonds	$BuyB_B$	$BuyB_B$				
Interest Bonds	$-Int_B^B$					$-Int_B^B$
Treasury borrows from CB	$Lend_{CB}^T$				$Lend_{CB}^T$	
Banks sell Bonds		$-SellB^P$	$SellB^P$			
CB buys from Banks		$-BuyB_{CB}^B$		$BuyB_{CB}^B$		
CB buys from Public			$BuyB_{CB}^P$	$BuyB_{CB}^P$		
Interest Bonds Public	$-Int_B^P$					$-Int_B^P$

Central Bank CB

	Asset		Liability		Net worth
	$Loans_{CB}^T$	$Bonds_B$	Reserves	Treasury	CB_E
Government Spend			Spend	-Spend	
Government Tax			-Tax	Tax	
Sell Treasury Bonds			$-BuyB_B$	$BuyB_B$	
Interest Bonds			Int_B^B	$-Int_B^B$	
Treasury borrows from CB	$Lend_{CB}^T$			$Lend_{CB}^T$	
CB buys from Banks		$BuyB_{CB}^B$	$BuyB_{CB}^B$		
CB buys from Public		$BuyB_{CB}^P$	$BuyB_{CB}^P$		
Interest Bonds Public			Int_B^P	$-Int_B^P$	

Bank

	Asset			Liability	Net worth
	Reserves	$Bonds_B$	Loans	Deposits	$Bank_e$
Government Spend	Spend			Spend	
Government Tax	-Tax			-Tax	
Sell Treasury Bonds	$-BuyB_B$	$BuyB_B$			
Interest Bonds	Int_B^B				Int_B^B
Banks sell Bonds		$-SellB^P$		$-SellB^P$	
Interest Bonds	Int_B^P			Int_B^P	
CB buys from Banks	$BuyB_{CB}^B$	$-BuyB_{CB}^B$			
CB buys from Public	$BuyB_{CB}^P$			$BuyB_{CB}^P$	

Public

	Asset		Liability	Net worth
	Deposits	$Bonds_P$	Loans	$Public_E$
Government Spend	Spend			Spend
Government Tax	-Tax			-Tax
Banks sell Bonds	$-SellB_P$	$SellB_P$		
Interest Bonds	Int_B^P			Int_B^P
CB buys from Banks				
CB buys from Public	$BuyB_{CB}^P$	$-BuyB_{CB}^P$		

Key	Flow Transaction Variable
$BuyB_B$	Treasury sells Bonds
$BuyB_{CB}^P$	Central Banks buys Bonds from Public
$BuyB_{CB}^B$	Central Bank buys Bonds from Banks
Int_B^P	Interest on Bonds paid to Public
Int_B^B	Interest on bonds paid to Banks
$Lend_{CB}^T$	Lending from Central Bank to Treasury
$SellB^P$	Sales of Bonds to Public
Spend	Government Spending
Tax	Government tax

Key	Stock or Account
$Bonds_B$	Treasury Bonds
$Bonds_P$	Publicly owned Bonds
$Bonds_{CB}$	Central Bank owned Bonds
Deposits	Deposit account of Public at a Bank
$Loans_{CB}^T$	Loans from Central Bank to Treasury
Reserves	Sales of Bonds to Public
Treasury	Government Spending

Note

1. Government Treasury spends and takes taxes; Deficit = Spend – Tax
2. Deficit is covered by selling Treasury Bonds
3. Treasury sells Treasury Bonds ($BuyB_B$) to Primary Dealer Banks and to Public
3. Banks sells Bonds (Treasury Bonds & Bank Bonds = $SellB^P$)
4. Central Bank buys Bonds in 'open market operations' ($BuyB_{CB}^P$ & $BuyB_{CB}^B$)
5. Treasury pays interest on Treasury Bonds (Int_B^B & Int_B^P)
6. Treasury borrows Interest from Central Bank ($Lend_{CB}^T$)

Figure A.1 Simplified Godley tables showing flows of money between the UK Treasury, the Central Bank, the Banking and Private Sectors

It is also worth noting that a transaction that affects both the Asset and the Liability/Equity sides of the Banking Sector's ledger creates money. When Central Banks buy Bonds from Banks only assets are affected, and so no new money is created, and the Treasury will no longer pay interest. Central Bank purchases of Bonds from the public do create money, however, since the sale credits both the public's deposit accounts at banks, and the reserve accounts of the banks at the Central Bank. It follows that the sale of bonds by the Banking Sector to the non-bank Public destroys money as the Public's deposit accounts fall, and their bond holdings rise. But that destroyed money was initially created by the deficit anyway. As long as the value of bonds sold by the Treasury is equal to or less than the deficit, the banks have enough reserves to buy them. In Figure A.1 Buy_T^B is taken from reserves to the Treasury. If all the bonds purchased by the banks from the Treasury were sold to the public, then the money created by the deficit would fall to zero.

The two questions posed at the head of this Appendix were 'How is money created, what are the transaction and where does the money end up?' A country with fiat money that issues its own currency can create money by running a deficit. The deficit creates net equity for the non-bank public, while interest on government bonds creates net equity for the banking sector. There is a kind of 'magic money' tree but one that is highly constrained. New money is created when you, as an individual or company in the public sector, borrow from a bank. The bank credits your deposit account at the bank electronically. The money didn't exist before. You can then withdraw it as cash or spend it electronically using, for example, a credit card or bank transfer as you wish. As you pay back your loan, as you must, the electronic money is deleted and hence no longer exists— it is destroyed. Banks can't lend money this way without limit because they are regulated. They are required to hold a certain amount of money as investment capital in their reserve account at a central bank. So, when a bank creates money, it doesn't actually create it out of thin air. In fact, it creates money out of assets of third parties, for example, a credit card company, a bank, and a central bank. If you pay for a coffee using a credit card, you are effectively giving an asset— your future repayment—and in return you get a credit card company IOU backed by the assets of a bank. All of these transactions can be captured in a stock/flow form as we see in Figure A.1 albeit in a very simplified form to illustrate the main points.

Finally, money can be created by quantitative easing as you will recall from Chapter 5. Quantitative easing is the electronic creation of money by a central bank buying assets on the open market. Typically, these are government bonds bought from pension funds and insurance companies when interest rates are low. Quantitative Easing increases the level of central bank money. But it is used only internally, by banks to pay each other. It also increases the level of commercial bank money as bank deposits which can be spent in the real economy.

The importance of these points is the impact on government policy regarding climate change. The fear of debt is holding back the crucial investment required to tackle the problems we face. From more renewable energy to strategies to ameliorate the damage that is being caused to our infrastructure and more importantly to the natural world we need governments to spend to initiate investment.

End Notes

1. For those readers who want to delve into even more detail see Berkeley, A., Tye, R., Wilson, N. 2021. An Accounting Model of the UK Exchequer. The Gower Initiative for Modern Money Studies at https://gimms.org.uk/2021/02/21/an-accounting-model-of-the-uk-exchequer/ The authors show that the ultimate source of 'moneyness' in the UK is the Consolidated Fund and not the Bank of England. The source of money is Parliament, and the Consolidated Fund is the Government's general bank account at the Bank of England. An opaque and complex pyramid of accounts is managed by the Debt Management Office. They comprise the National Loans Fund, The Consolidated Fund (the Exchequer) and the Contingencies Funds managed by the Treasury and at a more detailed level by the Office of the Paymaster General Government Banking Service. The Treasury determines the aggregate quantity of the net financial assets of the private sector, but the Bank of England monetary policy operations dictate the precise proportions in which these financial assets comprise reserves and government securities.
2. See Chapter 3, Blockley, D. 2022. Exchanging Obligations: Accounting for All Forms of Capital. Journal of Interdisciplinary Economics. 0(0). https://doi.org/10.1177/02601079221081717
3. The use of the word public can be confusing. There are two interpretations that must be held clearly. The public as the ordinary people—the community. The public sector as that part of an economy that is controlled by the state or government.

Glossary of Technical Economic Terms

Term	Definition
Account–chequing (current)	An account that includes deposit accounts demand accounts or transactional accounts. These accounts are very liquid and can be accessed using cheques, ATMs (automated teller machines), and electronic debits, among other methods. A chequing or current account usually allows unlimited deposits and many withdrawals *c.f.* savings accounts.
Account–current	A national current account records a country's transactions with the rest of the world—specifically its net trade in goods and services, its net earnings on cross-border investments, and its net transfer payments—over a defined period, such as a year or a quarter. The current account measures the country's net income. It is a measure of whether a country is a net borrower or lender. The deficit or surplus is usually captured as a percentage of GDP.
Account–demand	A bank account from which funds can be withdrawn without advance notice.
Account–savings	A savings account is a chequing (or current) account with some limitations on how often you can withdraw funds, but usually offering better interest rates.
Arbitrage	The simultaneous purchase and sale of an asset (stocks, commodities, and currencies) in different markets to exploit tiny differences in their prices. It takes advantage of the inevitable inefficiencies in markets.

Term	Definition
Asset quality	The likelihood of default of a loan or lease, combined with its marketability. The price of a loan or lease as determined by credit ratings agencies.
Automatic stabilizers	Financial 'shock absorbers'. Parts of fiscal policy that mitigate fluctuations without explicit government action. They temper the economy when it overheats and stimulate the economy when it slumps, without direct intervention.
Balance of payments	The balance of payments comprises a capital account and a current account. The capital account may sometimes be split into a financial account and a capital account. It captures the balance between national savings (that not spent on consumption) and investment. In a closed economy savings and investments are equal but usually they are decoupled and the difference between them is held in other countries. A current account surplus and capital account deficit means that money is flowing overseas.
Balance sheet	A statement setting out the assets, liabilities, and net worth of a financial entity at a particular point in time. In short it is a snapshot of what the entity owns and owes—the difference is the net worth which for a company could be shareholder equity. The UK national balance sheet in May 2022 shows rounded assets of £47bn, liabilities of £36bn creating a net worth of £11bn https://www.ons.gov.uk/economy/nationalaccounts/uksectoraccounts/datasets/preliminaryuknationalbalancesheetestimates
Bank of England Asset Purchasing Facility Fund	BAEPFF is a company that maintains a package of measures to support the UK economy as voted by the Bank of England Monetary Policy Committee. It was created in 2009 to implement the quantitative easing programme for the UK.
Banks	A bank is a company that receives deposits, makes loans and provides other financial services such as managing wealth, exchanging currencies and safe deposit boxes. Banks make a profit by charging more interest on loans they make than the interest they pay to people who save money with them. There are several different kinds of banks all regulated by national government or central bank. They include retail banks, commercial (or corporate) banks, and investment banks but the differences are often blurred.

Term	Definition
Banks–Central CB	Received opinion is that they are independent of government. But as they are owned by government, in actuality they are interdependent. Indeed, CBs must make any payments that the government instructs. So, unlike the rest of us, the government can spend whenever it wants. CBs do not deal directly with the general public. They are responsible for stability of currency, regulating capital and reserve requirements of banks, monetary policy, controlling inflation and the supply of money in a country. The major central banks include the Bank of England, the U.S. Federal Reserve Bank, and the European Central Bank.
Banks–commercial	Offer to their business clients services such as day-to-day business banking, credit, managing cash, real estate, employer, and trade finance.
Banks–Investment	Offer to corporate clients' services that maybe more complex such as underwriting and assisting with mergers and acquisitions. They often act as intermediaries for governments, large corporations, pension funds, and hedge funds.
Bank rate	The interest rate charged by a central bank for lending funds to commercial banks. Higher rates lead to higher lending rates by those banks.
Banks–retail	Banks that offer public services such as chequing (or current) and savings accounts, loans and mortgages, overdraft protection, credit card accounts and foreign currency exchanges.
Banks–Private Sector	Where most shares are held by the private individuals and corporations.
Banks–Public Sector	Where most shares are held by governments.
Borrowing–Public Sector Net Borrowing (PSNB)	The difference between government income and its current and capital expenditure—see also deficit.
Business–cycle	The business cycle of boom and bust is alternating bouts of a vicious cycle and a virtuous cycle. In a vicious cycle, output, employment, income, and sales decline and feedback into further declines (a domino effect) which spread across industries and regions. In a virtuous cycle the reverse happens—a recovery.
Capital	Anything that confers value or benefit. Examples include financial, natural, human, social, intellectual, cultural, and experiential.

Term	Definition
Capital account	A capital account captures the changes in national ownership of assets as the part of the balance of payments. It records all transactions made between a company and the rest of the world including imports and exports of goods, services, capital, and foreign aid.
Capital adequacy	The ratio of a bank's capital to its risk-weighted credit exposures. It is regulated to protect depositors and promote stability and efficiency.
Capital expenditure	Creates a lasting benefit or asset beyond the current year—not day to day expenditures.
Counter-cyclical	For example, stocks and shares that outperform during recessions and underperform in expansions.
Current expenditure–also revenue expenses	Government recurrent spending on goods and services in a current year. It includes social benefits, interest payments and other government department spending but excluding spending on capital assets.
Current receipts	Government income from taxes, interest, dividend, and rent.
Debt interest ceiling	The share of total public sector revenue spent on debt interest should not exceed a given threshold (say 10%) of that revenue at any time. The aim is that the overall debt is sustainable, taking account of the size, cost, and servicing.
Debt	The state of owing something—money, goods, or services. For most of us debt is serious. Government debt is not an issue—it is just money that the government has created and not taxed back because it is still of use in the economy.
Debt deflation	Recessions and depressions are due to the overall level of debt shrinking—deflating.
Debt Management Office UK	DMO is the UK Government's debt management agency that manages the sale of gilts and bills.
Debt–Public Sector Net Debt (PSND)	The amount of money the public sector owes to the private sector—including overseas.
Deficit	The difference between total expenditure and total revenue in public sector finances. It can refer to the current budget or to net borrowing when current and capital expenditure exceeds the current and capital income. See also Borrowing—PSNB.
Demand	Demand is the amount of goods and services that consumers are willing and able to buy over a given period—it is their desire and ability to pay.

Term	Definition
Diminishing returns	The law of diminishing returns states that the more capital you add with the same number of workers then the extra output gets smaller and smaller. Just as the satisfaction of eating, after a certain point, reduces.
Division of labour	Division of labour is the separation of tasks so people can specialise—it is a source of economic interdependence, but it overlooks hidden costs.
Double entry bookkeeping	A two-sided (debit and credit) way of accounting finances. Every entry to an account has a corresponding and opposite entry to a different account so that a transaction always affects at least two accounts and always equal total debits and total credits.
Dual Mandate	Coequal objectives of maximum employment and stability of prices (low inflation—usually around 2%).
Economics–classical	Classical economics flourished after Adam Smith in late 18th and early 19th centuries rejected mercantilism. It sees the economy as a self-regulating system governed by natural laws.
Economics–Keynesian	Keynesian economics is named after British economist John Maynard Keynes (1883–1946). It includes various macroeconomic theories and models about aggregate demand influencing economic output and inflation. Aggregate demand does not necessarily equal the capacity of production but is influenced by many variable factors.
Economics–neoclassical	Neoclassical economics is the study of the behaviour of entities (individuals, households, and organizations), as they use scarce resources to achieve their objectives. It assumes entities are capable, rational with stable preferences, have many objectives and limited resources.
Econometrics	Applying statistical and mathematical techniques to economics and, where possible, testing models.
Economy–command	A command or planned economy is one where production is publicly owned, and activity is under the control of a central authority (government), for example—prices, levels of production and allocations and distribution of resources. Levels of investment and consumption are also decided centrally often as a multi-year plan.
Economic cycle	The fluctuations of the economy between periods of expansion (growth) and contraction (recession).
Economy–free-market	A free-market economy is one where many private entities freely decide production and price levels. The prices are self-regulated by buyers and sellers negotiating

Term	Definition
	in an open market. The supply and demand of goods and services are free from any intervention by a government or other authority.
Economy–mixed	A mixed economy is a combination of free and command economies.
Elasticity	The sensitivity of demand for a good or service to changes in prices or other market variables.
Entity	A financial entity is an individual, group, organisation, public authority, or government engaged in financial activity.
Expenditure	Government current expenditure includes interest payments, benefits (e.g., pensions), procurement of goods and services (e.g., defence, construction), wages, transfers to local governments, and subsidies. Total expenditure includes depreciation and net investments.
Exchange rates	The relative value of one currency compared to that of another. The higher the rate the less expensive are imports and exports are more expensive in foreign markets. Rates are influenced by interest rates and inflation.
Externality	An externality is a cost or benefit not felt by an entity—a contextual side-effect and often environmental. Examples of positive externalities are education and research and development. Negative externalities pollution and climate change. Externalities interfere with Adam Smith's invisible hand. For example, an educated work force tends to be more productive, law abiding with more civic engagement.
Fiat currency	Government issued currency not backed by a commodity such as gold.
Financial innovation	In recent times the focus has been on derivatives markets and securitisation, seen as driving the growth of capital market activities.
Financial Instability	Herman Minsky, see https://www.levyinstitute.org/pubs/wp74.pdf said '*capitalist economies exhibit inflations and debt deflations which seem to have the potential to spin out of control...Government interventions aimed to contain the deterioration seem to have been inept in some of the historical crises*'. He argued that this falsifies the idea that the economy can best be understood by assuming that it is constantly an equilibrium seeking and sustaining system.

Term	Definition
Financial instrument	A tradeable asset, including a package of capital, such as cash, a contractual right to deliver or receive cash or another type of financial instrument, or evidence of ownership.
Financial repression	Holding interest rates below the rate of inflation. Public debt is eroded over time, but the assets of private sector creditors drop in value.
Financial Sector Assessment Programs	Careful assessments of macroprudential vulnerabilities.
Fiscal deficit	The difference between government expenditures and revenue usually quoted as a percentage of GDP.
Fiscal headroom	A notional measure of how much the government can borrow without damaging the economy. It is the difference between the limit set by a fiscal rule and projected levels of borrowing.
Fiscal policy	Government actions to set taxes and spending to steer the economy.
Fiscal rules	Restrictions on fiscal policy set by the government to constrain its own decisions on spending and taxes. For example, they might require that the deficit (the difference between government spending and revenues) stays below a certain level.
Fiscal space	flexibility in a government's budget so that it can resource a purpose without threatening the stability and sustainability of its economy. It is often measured as the ratio of total debt to tax revenue.
Foreign direct investment flows (FDI)	Investments made by companies or individuals of one country into another different country.
Fractional reserve banking	Where banks must hold a portion of the money deposited with them as reserves. The banks use customer deposits to make new loans and return interest.
GDP	The nominal Gross Domestic Product is an estimate of national economic activity. It comprises consumption C, business investment I, government spending S, and net exports, NX, i.e., GDP = C + I + S + NX. It can also be estimated by the total income paid to everyone in a country, i.e., wages, rent, interest, and profit on capital with adjustments to add indirect business taxes, depreciation, and net foreign income. Real GDP is nominal GDP adjusted for inflation with respect to a base year. USA Quarter 1 in 2022, C = 68%, I = 18%, G = 18%, NX = –4% https://fred.stlouisfed.org/release/tables?rid=53&eid=12998#snid=12999

Term	Definition
GNP	Gross National Product is an estimate of all products and services of a given period through production owned by a nation's residents. It is GDP plus foreign investment income of residents minus foreign residents' income that has been earned within the country.
Gilts	A government security or bond issued by a national treasury.
Goodhart's Law	Goodhart's law says that a measure should not be a target. The reason is that unintended consequences will interfere with any controls. For example, a focus on money changes the connection between money and inflation because the velocity of money can fluctuate.
Headroom	The difference between the required and available cash. See also fiscal headroom.
Helicopter money	See Monetary Financing.
Inflation	Inflation is the rate of increase in prices over a range of goods and services over a given period. Typically interest rates are raised to reduce inflation because companies and individuals then have to repay more for their loans—money that they might otherwise have spent on goods and services.
Interest	The amount a lender charges a borrower as a percentage of the amount loaned. When noted annually it is the annual percentage rate (APR).
Interest rates	There are many types of interest rates depending on the type of loan. For example, savings rates will depend on bank rate plus other factors such as economic uncertainty such as recessions and economic shocks. Mortgages, credit cards and payday loans are all important in day-to-day living.
Investment	The placing of money, effort, or time into something to advantage such as a profit. It includes the buying of stocks (as shares or equities), bonds, mutual funds, and commodities such as precious metal, land, or equipment.
Invisible hand	The invisible hand is a metaphor for the hidden pressures that cause changes in the economy of a free market. Self-interest and individual drive the best interests of society in a dynamic of supply and demand. It is a natural flow of trade. For example, no-one tells a factory how much to produce—it is a matter of individual or group judgement—everyone is acting in their own self-interest. The invisible hand seems to say that greed is good—but that was not what Smith intended.

Term	**Definition**
Laffer Curve	As tax rates increase, some have a reduced incentives to work and make investments because they make less money from them. Above a certain tipping point rate, total revenue from taxes goes down because people slow down their economic activity. The tipping point is difficult to determine.
LIBOR	London Inter-Bank Offered Rate is an average interest rate (in several currencies over time periods ranging from overnight to one year) based on estimates by the leading banks on what each would be charged if it borrowed from other banks.
Liquidity	The ease with which an asset can be converted into cash without affecting its market price or a financial instrument can be exchanged for goods and services.
Liquidity trap	A decline in economic activity, through low confidence and unwillingness by companies to invest. In a recession firms and consumers may have high levels of debt which they pay off and save more and spend less.
Loan	A sum of money that is expected to be paid back with interest.
Loanable funds	The total of all of the money that people, and financial entities have decided to save and then lend as an investment rather than consume personally.
Loan–payday	Short-term, very-high-interest loans.
Macroeconomics	Macroeconomics is the study of the economic behaviour of nations, governments, and whole industries.
Macroprudential indicators or financial soundness indicators	These include GDP growth, inflation, volatility of interest and exchange rates, capital adequacy, asset quality, earnings, liquidity, and sensitivity to market risk, see https://www.imf.org/external/pubs/ft/op/192/OP192.pdf
Macroprudential policy	To promote the stability of the whole financial system—not just individual institutions—the safety and soundness of the broad financial system and payments mechanism. They are intended to prevent substantial disruptions in credit and other vital financial services necessary for stable economic growth.
Macroprudential tools	May be structural or cyclical. Structural build resilience in the business cycle such as limits on loans, debt to income that protect borrowers from too much debt. Cyclical increase resilience against economic downturns such as buffers required of banks to absorb losses. They are also tools to stabilise markets and securitisation.

Term	Definition
Macroprudential vulnerabilities	A lack of transparency in the markets for example derivatives and concentration in a few institutions to undermine robustness of liquidity. Regulatory arbitrage; the under-pricing of risk on new instruments; the overestimation of their liquidity; the opaqueness of risk resulting from interconnections in the financial system; the danger of risk concentrations; the overloading of payment and settlement systems, reflecting a sharply higher volume of transactions; the potential for increased market volatility; and stronger growth in overall debt.
Magic money tree	Money can be plucked out of thin air. There is no gold standard. Money is created on faith. If too much is created too then we get inflation. Money has no innate value—it is a marker of trust between a lender and a borrower and backed by a third party.
Market–command	A command economy is one where a central authority (government) decides how and what to produce at what price. People's earnings do not depend on what they do—rather they are centrally regulated. Profit is not the driving force.
Market–failure	Markets are deemed to have failed when the distribution of goods and services in a free market are inefficient. This happens when incentives for individuals and groups do not align. In other words, decisions that are good for individuals are wrong for the group. For example, supply does not equal demand or an externality like global warming has unintended side effects on the market.
Market–free	A free-market or *laissez-faire* economy is one where individuals and businesses are free to pursue their own economic interests. They buy and sell goods in competition and the market determines the fair price. Intervention by authority or government is minimal.
Market economy	A market economy is one where investments, production and distribution of goods and services to consumers are actuated by supply and demand with no restrictions or controls.
Market maker	A dealer in securities or other assets.
Mercantilism	Mercantilism is an economic system that replaced the feudal system in the 16th to 18th centuries. It assumes a zero-sum game—one man's gain is another man's loss. The power of a nation could be increased by increasing exports and minimising imports and consequently

Term	Definition
	supplies and markets were protected. For example, the English Sugar Act of 1764 raised duties on foreign refined sugar and molasses imported by the colonies, to create a monopoly for British sugar growers in the West Indies.
Microeconomics	Microeconomics is the study of the choices of individuals as buyers, sellers, owners, as they respond to incentives, prices, resources, and production methods.
Modern Monetary Theory	MMT maintains that countries that issue their own fiat currency, like the US, UK, Japan, and Canada, are not operationally constrained in the same way as currency users such as individuals, households, companies, and countries with non-fiat currency.
Monetary Financing	A direct transfer of money from a central bank for a government to spend. It usually involves the direct purchasing of new government debt (bonds) by a central bank. It can lead to less discipline in government borrowing and inflation since it increases the supply of money. The risk is low when demand is weak and there is a need for government investment. This policy is distinct from quantitative easing (QE) where the central bank purchases bonds in the existing secondary market. It is called by some '*helicopter money*'.
Money–broad	M3 and M4
Money–definition	The definitions of the four types of money as M0, M1, M2, M3 and M4 vary slightly across different countries. The principal features are as described in this glossary.
Money–M0	M0 is the monetary base, the most liquid form of money, and an obligation of a central bank. It includes all currency as coins and banknotes either in general circulation or as the deposits of commercial banks held in a central bank's reserves.
Money–M1	M1 is an obligation of a commercial bank. It includes M0 plus travellers' cheques, demand, savings, and other liquid deposits such as chequing (or current) accounts through debit cards. It does not include financial assets such as bonds.
Money–M2	M2 is an obligation of a commercial bank and is M1 plus savings deposits maturing in less than two years. Note it is not a claim on the state or central bank but an IOU from a commercial bank.

Term	Definition
Money–M3	M3 includes M2 plus longer-term deposits, repurchase agreements, debt securities up to two years and money market shares.
Money–M4	M4 is the least liquid form of money—aggregate money. It includes M3 plus all other deposits.
Money–market	Money markets are trade between financial institutions and companies in very short-term investments, such as bonds, bills, notes, and other securities. It is the buying of a large quantity of debt to receive interest. The trade includes overnight swaps of large sums of money between banks and government.
Money–narrow	Narrow money (M0 in the UK and M1 in USA) is a category of money supply that includes all physical money such as coins and currency, demand deposits, and other liquid assets held by the central bank.
Money–supply	The money supply comes in four different forms called M0, M1, M2, M3. It is increased when banks lend, central banks reduce reserve ratios for bank lending, or they buy government securities or invoke quantitative easing. Increasing the money supply can cause inflation but depends on velocity of circulation of money.
Money–velocity of circulation	The amount of money circulated in the economy during a given period of time usually measured by dividing GDP by total money supply.
Multiplier	An economic factor that, when increased or changed, causes increases or changes in many other related economic variables. The multiplier causes gains in total output to be greater than the change in spending that caused it.
Net worth	Excess of assets over liabilities. For the public sector the excess of the government's total financial and fixed assets over its debt and other liabilities expressed as a share of GDP.
Nominal	Nominal means something that is not adjusted for inflation.
Non-financial account	A record of current expenditure, current revenue, and capital transactions.
Open market operations	The trading of government securities by a central bank to regulate money supply. The central bank sells securities to reduce the money supply and increase interest rates. It buys to increase the money supply to reduce interest rates and help growth. The public is not involved.

Term	Definition
Opportunity cost	Opportunity cost is not just what you pay but also what you give up.
Primary dealer	A bank or other financial entity that has been approved to trade securities with a government. In many countries, they are the only ones who can bid for newly issued government securities.
Prisoner's dilemma	The prisoner's dilemma occurs when an individual's best choice is not the best choice for the whole group. For example, in an arms race, the best option would be for neither state to invest in weapons. Economic incentives can induce behaviour that makes everyone collectively worse off, while individually avoiding choices that would make them all collectively better off if choice could be made collaboratively.
Procyclical	Consistent with the changes in an economic cycle.
Public sector	That part of the economy controlled by the state.
Public sector–net cash requirement (PSNCR)	The amount the government needs to raise to meet the shortfall in net borrowing. Public sector net debt (PSND) is roughly the sum of all the historical public sector net cash requirements.
Purchasing power	Purchasing power is the amount of money a person or group has available to spend.
Quantitative easing (QE)	Central bank buying of existing government bonds in the secondary market when interest rates are already low. This has the effect of raising the price of bonds and lowering interest rates. Credit becomes cheaper and encourages spending and investment. Banks are more likely to offer loans.
Rational ignorance	Where a person decides to act without having all of the information that could be obtained. Sellers may increase the complexity of a decision so that the difference in value between products is less than the effort required to differentiate them. It becomes more rational for a consumer to just take his chances on whichever of the two is more convenient and available. Politicians may increase the number of issues that a voter needs to consider. People do not have the time to research every aspect and so choose based on prior experience rather than a full analysis of options.

Term	Definition
$r > g$	If the interest rate r is greater than g the rate of growth of the economy, then savers benefit by getting a larger share of national income. Borrowers, producers, and speculators are taxed to subsidize savers and owners of monetary assets. Clearly when $g > r$ the opposite is the case. However, g and r are not independent, so they tell us little about the sustainability of debt.
Real	Real means something that is adjusted for inflation.
Recession	A recession is a contraction of the economy, as measured by GDP, for more than 6 months.
Rents	Rents are revenues over and above what is possible to earn in a competitive market—for example, tariffs on foreign goods over domestic goods.
Repo	A sale and repurchase agreement. If the repurchase price is 10% higher than purchase price the repo rate is 10%.
Reserve	Money held and not lent out.
Resource curse	Nations with rich natural resources (such as petroleum or minerals) squander them, maybe through corruption, foreign ownership, or authoritarian regimes—resulting in poor economic development.
Revenue	The total amount of income. Government revenue includes taxes and fees, interest, and loan payments minus subsidies. It also includes funds raised by issuing bonds to banks, investors, and foreign currency markets.
Secondary market	One in which investors trade securities they already own. The stock market is an example though when first issued they are sold on the primary market.
Security	A readily interchangeable asset of monetary value. It can be a stock, a bond, or a right to ownership. The three main types are debt securities, equity securities, and derivative securities.
Securities–debt	Bonds and certificates of deposit, as a rule, require the holder to make the regular interest payments, as well as repayment of the principal amount alongside any other stipulated contractual rights. Such securities are usually issued for a fixed term, and, in the end, the issuer redeems them.
Securities–equity	represent ownership interest held by shareholders in a company. In other words, it is an investment in an organization's equity stock to become a shareholder of the organization.

Term	Definition
Securities–derivative	A derivative security is a financial instrument whose value depends upon the value of another asset. The main types of derivatives are futures (buy/sell at some predetermined prices at future specified time), forwards (as futures but price to be agreed at end of contract), options (holder has right but not the obligation to buy/sell), and swaps (agreement to exchange).
Shrinkflation	Shrinkflation is reducing the amount of a good but keeping price the same, for example, putting less chocolate in a bar.
Stagflation	Stagflation is high unemployment and high inflation.
Stress test	A computer simulation to assess the financial health of organisations such as banks, building societies and insurers against possible future scenarios.
Subprime mortgage	A mortgage issued to borrowers who have a low credit rating. Because the risk of default is higher interest rates are normally higher than conventional mortgages and can be changed.
Supply	Supply is the total amount of a good or service available to consumers at a given time and price.
Tax	A compulsory financial charge on an individual or legal entity by a government. Direct taxes are taxes on income, profits, and wealth. Indirect taxes are taxes on expenditure (e.g., VAT).
Tax–distortion	A harmful side effect of tax, for example, when people do things for tax reasons that otherwise make no economic sense. Where a choice is motivated by tax and not just economic costs and benefits. Minimizing one's tax burden by various means such as giving to charity or your spouse, timing when to receive income, etc. Whilst making sense for individuals these actions are a real economic waste for countries.
Tax–incentive	Exclusions, exemptions, deductions, refunds, and credits from taxes owed. Businesses receive them to invest into their businesses, to support the environment or minorities or disadvantaged business owners.
Tax–instrument	The way taxes are raised. Also known as tax structure.
Tax–progressive	When tax rates are higher for higher levels of income.
Tax–regressive	When tax rates decrease for higher levels of income. It imposes a greater tax burden on the poor than on the rich.

Term	Definition
Tinbergen rule	To achieve several macroeconomic targets a policy maker must be able to control an equal number of macroeconomic instruments.
Tragedy of the commons	The tragedy of the commons is an example of the prisoner's dilemma. It may be in everyone's collective advantage to conserve and reinvest in the propagation of a common pool natural resource in order to be able to continue consuming it, but each individual always has an incentive to instead consume as much as possible as quickly as possible, which then depletes the resource. Finding some way to co-operate would clearly make everyone better off here. The tragedy is that if everyone acts in own self-interest then sometimes the system as a whole degrades and loses value.
Treasury	Is a ministerial economics and finance department that maintains control over public spending, setting the direction of economic policy and achieving economic growth.
Treasuries	USA government bonds.
Treasury Bills	Have the shortest range of maturity in terms of weeks.
Treasury Bonds	Have the longest maturity of 30 years.
Treasury Notes	Have the middle range of maturity from 2 to 10 years.
Treasury Securities	Can be purchased from and are fully backed by USA government. They are of three types—T-Bills, T-Bonds and T-Notes according to the length of maturity. They are vulnerable to changes in inflation and interest rate changes.
Ways and means	Methods and resources to accomplish something—for raising the necessary revenues for the expenses of a nation or state, the taxes, and other charges to fund government spending.

Index

Praise

Name	Quote
Professor Jim W. Hall Professor of Climate and Environmental Risks Oxford Programme for Sustainable Infrastructure Systems University of Oxford, UK	*David Blockley explains how we should think about and act to address the extraordinary challenge of climate change. He does so with humility and humanity, drawing upon wisdom from across the ages and around the world.*
Patrick Godfrey Emeritus Professor University of Bristol, Founding Coach of International Council of System Engineering's Technical Leadership Institute.	*This inspiring book sets out the unintended consequences of our impact on the natural world. It shows that, it is not yet too late to learn from nature, how to face up to our obligations to each other and the future.*
Priyan Dias Emeritus Professor in Civil Engineering, University of Moratuwa. Immediate Past President, National Academy of Sciences of Sri Lanka.	*Embark on an enlightening journey through the exceptional perspective-bridging career of engineering academic Professor David Blockley. From engineering to ethics, and innovation to economics, Blockley's expertise traverses a spectrum of fields, interconnecting them with finesse.*